John t
johnpa   r.net

Millimetre-wave Optics, Devices and Systems

Millimetre-wave Optics, Devices and Systems

J C G Lesurf

University of St Andrews, UK

Adam Hilger, Bristol and New York

© IOP Publishing Ltd 1990

All rights reserved. No part of this publication may be reproduced, stored in a retrieval system or transmitted in any form or by any means, electronic, mechanical, photocopying, recording or otherwise, without the prior permission of the publisher. Multiple copying is only permitted under the terms of the agreement between the Committee of Vice-Chancellors and Principals and the Copyright Licensing Agency.

British Library Cataloguing in Publication Data

Lesurf., J. C. G.
  Millimetre-wave optics, devices and systems
  I. Title
  621.381'3

  ISBN 0-85274-129-4

Library of Congress Cataloging-in-Publication Data

Lesurf, J. C. G. (James Christopher George)
  Millimetre-wave optics, devices, and systems / J.C.G. Lesurf.
     p.   cm.
  Includes bibliographical references.
  ISBN 0-85274-129-4
  1. Millimeter waves. 2. Millimeter wave devices. 3. Optical instruments.   I. Title.
  TK7876.5.L47   1990
  621.36—dc20

Consultant Editor: **Mr A E Bailey**

Published under the Adam Hilger Imprint by IOP Publishing Ltd
Techno House, Redcliffe Way, Bristol BS1 6NX, England
335 East 45th Street, New York, NY 10017-3483, USA

Typeset by MCS, Salisbury, England
Printed in Great Britain by Eastern Press Ltd, Reading

Dedication and thanks

To Chris, who has put up with my sore head whilst writing this book, and to Derek Martin, who still knows a lot more than me about millimetre waves.

Dedication and thanks

Contents

Preface		xi
1	**Signal transmission, modes and Gaussian beams**	
	1.1 Waveguides and general mode properties	1
	1.2 Gaussian beams in free space	6
2	**Beam coupling, lenses and mirrors**	
	2.1 Coupling between Gaussian beams	11
	2.2 Lens and beam coupling	15
	2.3 Dielectric materials and surface reflections	18
	2.4 Mirrors and beam coupling	24
3	**Multi-mode beams and apertures**	
	3.1 Gaussian beams and diffraction	29
	3.2 Apertures and beam truncation	34
	3.3 Beam offsets and measuring beam patterns	39
4	**Antennas and feed systems**	
	4.1 Basic coherent antenna theory	44
	4.2 Feed horns	50
	4.3 The scalar feed	52
	4.4 'Vivaldi' and other E-plane antennas	56
	4.5 Behaviour of a Cassegrain system	58
5	**Transmission lines, impedance matching and signal reflections**	
	5.1 Transmission lines and terminating loads	65
	5.2 Semi-reflectors, dielectric slabs and mismatched lines	72

6 Bolometric signal detection
- 6.1 Basic properties of thermal detectors — 77
- 6.2 Types of bolometer — 86
- 6.3 Phase sensitive detection and background subtraction — 91

7 Mixers and heterodyne detection
- 7.1 Signal mixing in diodes — 97
- 7.2 Bulk mixers — 106
- 7.3 Signal-to-noise properties of heterodyne receivers — 108

8 Filters and resonators
- 8.1 Semi-reflecting sheets — 115
- 8.2 Metallic mesh — 119
- 8.3 Fabry–Perot and resonant filters — 123
- 8.4 Refocusing and 'ring' resonators — 129
- 8.5 Absorption filters — 133

9 The Martin–Puplett interferometer
- 9.1 The two beam (Michelson) interferometer — 130
- 9.2 Basic properties of the Martin–Puplett interferometer — 140
- 9.3 Gaussian beam mode analysis of the Martin–Puplett interferometer — 147
- 9.4 The limitations of fast Fourier transform spectroscopy — 150

10 The design of optical circuits
- 10.1 Circuit symbols and design — 156
- 10.2 Polarization states and matrix algebra — 161
- 10.3 An optical impedance measurement circuit — 166
- 10.4 Physical size and circuit bandwidth — 172

11 Oscillators and signal sources
- 11.1 Thermal sources — 181
- 11.2 Vacuum state coherent oscillators — 185
- 11.3 Solid state oscillators — 190
- 11.4 Solid state negative-resistance devices — 197

12 Frequency control loops and diplexers
- 12.1 Local oscillator control — 209
- 12.2 Gaussian analysis of Fabry–Perot 'ring' diplexers — 221
- 12.3 The Martin–Puplett interferometer as a diplexer — 224

Contents

    12.4 Gaussian analysis of the Martin–Puplett
           interferometer as a diplexer   227

Appendices
   1 Beam normalization and mode polynomials   233
   2 Circuit symbols and matrices   237

Further reading and selected references   **241**

Index   **247**

Preface

This book is primarily intended as a general introduction to the practical and theoretical methods used for the design of millimetre-wave optical instruments and devices. The emphasis is on providing a clear description of the underlying physics rather than attempting to discuss in great detail the minutiae of a wide range of devices and techniques. I hope that it will prove useful as a course text or reference book for advanced undergraduate and postgraduate students and as a 'first port of call' for physicists and engineers wishing to work on millimetre-wave systems.

What is the millimetre-wave region of the electromagnetic spectrum? In what ways do the methods used for millimetre-wave systems differ from those employed for visible light or for radio waves? The most obvious answer to the first question is to define millimetre waves as those which have a wavelength around a millimetre. Yet does this mean we should include a wide range of frequencies from 30 GHz to 3000 GHz? Or restrict our attention to a narrow range centred on 299 GHz? In practice the millimetre-wave region is more usefully defined as the range of wavelengths where the techniques appropriate for one millimetre wavelength continue to work well.

Until just a few years ago, millimetre-wave systems and measurements were largely confined to areas of laboratory or 'pure' science. More recently there has been a dramatic growth in the use of signal frequencies above 75 GHz for commercial, industrial and military applications. This has not, however, been matched by a similar growth in interest in the frequency range above a few hundred GHz.

One of the aims of this book is to outline for those working in this 'growth area' some of the optical methods which can be used to obtain levels of performance similar to—or better than—those obtainable using conventional waveguides in microwave systems. As a result, the examples and explanations have been chosen on the assumption that the reader is mainly interested in instruments and methods for the 75–500 GHz range where optical and waveguide items are often combined. It should be noted, however, that the techniques described in this book can easily be applied over a far wider range of signal frequencies.

Millimetre-wave systems have a great deal in common with microwave and radio systems. The book therefore also provides a simple introduction to some practical techniques like heterodyne reception and the use of transmission line methods which are likely to be useful to anyone who does not already have a background in these subjects. In order to keep the text brief, a number of significant related subjects have been omitted as not being directly appropriate to most of those who will read this book. The primary example of a deliberate omission has been the use of lasers as coherent signal sources.

Most millimetre-wave lasers are only useful in a specialized laboratory situation and at frequencies above a few hundred GHz. In general, commercial, industrial and military systems tend to employ vacuum or solid state coherent sources of the forms described in chapter 11. Furthermore, there is a plethora of books on lasers. Hence it has been decided to concentrate the text on those areas of more direct interest which have received little or no coverage elsewhere. In particular, there is an emphasis upon the use of Gaussian beam mode methods in understanding and designing millimetre-wave systems.

In many ways, the properties of Gaussian modes have a role in millimetre-wave optical systems similar to the conventional waveguide in microwave systems. Gaussian methods provide us with a way of coping with many of the problems which arise when we wish to devise systems which use optics in place of waveguides. As a consequence it is a topic which is of special relevance at signal frequencies where waveguides work poorly.

Some material, for example the sections on transmission lines, is presented in a slightly different way than in texts concerned with conventional microwaves. This is in order to bring out the common features of systems based upon signal transmission using free-space beams and waveguides.

An alternative definition of millimetre waves might be to say that they occupy the spectral region 'between microwaves and infrared'. In some ways it would be fair to describe millimetre-wave methods as 'part electronics, part optics', but in fact those methods reveal that both electronics and optics are portions of a greater whole. Traditionally, we tend to regard electronics and optics as being distinct subjects. In electronics, signals are sent in the form of voltages and currents along wires. In optics, signals are transmitted as beams of light. Electronic circuits use resistors, capacitors, transistors, etc, whilst optics uses lenses, mirrors and lasers.

A common way of introducing electronics at school has been to make use of a 'water-pipe' analogy. The flow of electrons along a wire is likened to

Preface

that of water in the pipe and the electrical voltage is likened to the pressure. The analogy is a useful one, but it can be misleading.

When we switch on a torch we expect the light to come on with no obvious delay. If the bulb does not light up after a minute or two we suspect that the battery is flat or the torch faulty. Yet if we calculate the mean velocity of the electrons along the wires we find that they take some minutes to go a few centimetres from the switch to the bulb! Clearly the signal that the switch is closed moves much faster than the electrons. So what tells the bulb that the switch has been closed?

The two sides of the open switch have different electronic potentials, i.e. the charge densities inside the metal wires are different. As the switch closes the electrons in the wires experience a force as the charges are brought together. This moves them slightly along the wire, changing the local charge density—and hence the potential—changing the forces on other electrons nearby. The change in charge density (and hence potential) ripples along the wire far faster than the actual electrons.

The signal from switch to bulb is an electromagnetic wave which transmits power along the wire. The velocity of light in air is far higher than in any normal metal. Hence the electromagnetic fields outside the wire change much more rapidly than those inside and almost all the signal power is carried along just above the surface of the metal.

The wire actually behaves as a waveguide, guiding the signal from switch to bulb at a velocity almost as high as that of light in a vacuum! The electrons also move along the wire, but the power carried by their kinetic energy is far smaller than that carried by the electromagnetic wave.

Light emitted by a bulb (or by a laser, a fluorescent tube, etc) carries energy away from an electron which has taken part in an interaction or energy transition. In order to detect the light, or any other electromagnetic radiation, we have to arrange for it to interact with a charged particle or body.

Electronics and optics require energy to be communicated between charges by an electromagnetic interaction. The division between electronics and optics can therefore be seen to be one of convention and technique. Millimetre-wave methods exploit the underlying unity of electronics and optics by borrowing whatever arguments or techniques best suit the task.

Whilst the main purpose of this book is to present information on millimetre-wave instruments and devices I hope also to take the opportunity to point out the underlying physics that links optics and electronics in an area of the spectrum where the two become indistinguishable. In this way I may perhaps produce a volume of equal interest to engineers and physicists.

Jim Lesurf
May 1989

Chapter 1
Signal Transmission, Modes and Gaussian Beams

1.1 Waveguides and General Mode Properties

Conventional electronic instruments use wires to send signals from place to place. As the signal frequency rises, it becomes increasingly difficult to make wired systems which work well. This is because—as was mentioned in the Preface—the actual signal is transmitted as an electromagnetic field which moves along just outside the wire. Some of this field can, therefore, be radiated away or coupled onto any other nearby wires.

Changes in the potential of a wire can be detected as a varying force on nearby charges. Similarly, changes in current alter the surrounding magnetic field and may induce currents on other wires. These effects are normally dealt with in electronics as 'stray' capacitances and inductances. They mean that some of the power that we wish to transmit via a wire may fail to arrive at its intended destination because it has been diverted elsewhere. It may also mean that some of the variations in potential and current arriving at the signal destination are the result of unwanted signals coupled onto the wire.

In many cases the signal wavelengths are far larger than the lengths of the transmitting wires and we can think of variations in current and charge as being uniform along the wire. If the frequency is increased sufficiently (or the wire extended), this assumption ceases to be reliable. Then the potential and current may be seen to vary along the length of the wire, i.e. there is a noticeable electric and magnetic field variation *along* the wire. Both of these fields vary periodically in time at the signal frequency. Now the wire will act as an *antenna*, radiating some signal power into the surrounding space.

As a result of these effects the efficiency of signal coupling along a wire tends to fall as the frequency rises. Various measures can be adopted to try to counteract these problems. One of the most useful is to replace the wires with a *metallic waveguide*. Although mainly used at microwave frequencies the metal waveguide is worth discussing in some detail here as it is used in many millimetre-wave systems and components. Many of the properties of waves propagating in waveguides of this type also turn out to be applicable in general to other forms of waveguide and to beams in space.

If we regard a normal wire as a length of metal surrounded by space, we can think of a metal waveguide as a length of space surrounded by metal. As with the wire, signal power is transmitted as an electromagnetic wave which moves along in the space outside the metal, i.e. the wave moves down the hole in the centre of the guide.

The most common form of waveguide is rectangular in cross section. Consider figure 1.1 which represents a rectangular guide.

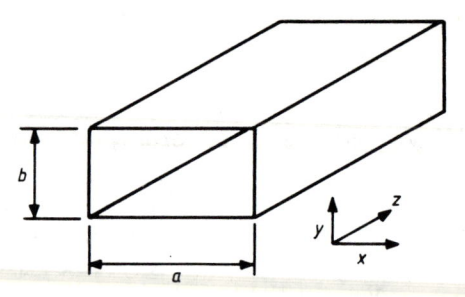

Figure 1.1 Short piece of rectangular waveguide.

In order to determine the field pattern inside the waveguide, we have to find the appropriate solution for Maxwell's equations. We need to know at least some of the field components at some point within the guide. How can we do this for a rectangular metal waveguide?

A rectangular waveguide can be regarded simply as a set of four mirrors placed so as to form a long metal box with open ends. With this in mind we can understand its behaviour by considering what happens when an electromagnetic wave is incident upon a metal surface.

An incident wave sets up a current in a thin layer at the surface. If the wave is plane parallel moving perpendicular to the surface, then the current distribution is uniform over the surface. Hence there will be no net accumulation of charge at any point on the surface. No potential differences are produced and the total electric field along the surface must remain zero everywhere.

The velocity of light inside a metal is generally far lower than in free space, i.e. the refractive index is very high. Hence very little of the incident wave penetrates into the metal. In order for the incident energy not to vanish mysteriously it must be reflected. As the total electric field along the surface must be zero, it follows that the incident and reflected waves have electric field components parallel to the surface which are equal and opposite.

$\omega = 2\pi\nu$ $j = \sqrt{-1}$ $\mu_0 =$ permeability $\varepsilon_0 =$ permittivity
$\tilde{H}_0 =$ max. mag field of guide   Power $\propto |H_0|^2$
(a measure of) ($H = \frac{B}{\mu}$?)

Signal Transmission, Modes and Gaussian Beams

A wave incident upon a metal surface at an angle can be regarded as a combination of *two* waves arriving simultaneously—one perpendicular to the surface, and the other parallel to the surface. The effect of the perpendicular wave is as described. The behaviour of the wave moving parallel to the surface depends upon the orientation of its electric field with respect to the metal surface. However, any current flow set up in the metal will be such that there can be no electric field at the surface perpendicular to the direction of propagation of the wave.

For the case of a set of mirrors arranged to form a rectangular guide we find that we cannot produce an electric field at a guide wall which is both parallel to the wall and perpendicular to the axis of propagation along the guide.

We can consider the signal power propagating along the waveguide as being guided by repeated reflection between opposite walls. Although this is a simple model of what takes place, it serves to introduce some of the basic properties of signal transmission using a metallic waveguide.

The first property that we may expect is that the electric field component parallel to a metal wall should be zero at the wall surface. We can also expect the rate of progress of a signal along the guide to be less than for a wave in free space. This is because the signal does not simply move parallel to the guide axis but takes a longer path, 'tacking' back and forth across the guide like a yacht travelling upwind.

From knowing that particular electric field components must be zero at the walls we can find the appropriate solutions for Maxwell's equations and determine the actual field pattern within the guide.

For a rectangular waveguide the solutions fall into two general classes, referred to as *transverse electric* (TE) and *transverse magnetic* (TM) waves. The electric field of a TE wave is everywhere perpendicular to the guide axis of propagation. Similarly, the magnetic field of a TM wave is everywhere perpendicular to the guide axis.

It is useful to examine the properties of these waves in a little detail as this will enable us to see some of the basic properties which are common to all wave-guiding systems—even those consisting of a beam in space. Most metallic waveguide systems use TE waves; hence we can use them for the purpose of example.

The solutions of Maxwell's equations for TE waves in a rectangular metal waveguide are of the general form

$$E_x = \frac{j\omega\mu_0 H_0 n\pi}{k_c^2 b} \cos\left(\frac{m\pi x}{a}\right) \sin\left(\frac{n\pi y}{b}\right) \exp[j(\omega t - \beta z)] \qquad (1.1)$$

$$E_y = \frac{-j\omega\mu_0 H_0 m\pi}{k_c^2 a} \sin\left(\frac{m\pi x}{a}\right) \cos\left(\frac{n\pi y}{b}\right) \exp[j(\omega t - \beta z)] \qquad (1.2)$$

$$E_z = 0 \qquad (1.3)$$

where

$$k_c^2 = \left(\frac{m\pi}{a}\right)^2 + \left(\frac{m\pi}{b}\right)^2 \qquad (1.4)$$

and β = propagation const.

$$\beta^2 = \omega^2 \mu_0 \varepsilon_0 - k_c^2. \qquad (1.5)$$

The Cartesian coordinates are defined as shown in figure 1.1 with the z axis parallel to the guide propagation axis. The guide width and height are a and b; ε_0 and μ_0 are the permittivity and permeability of free space; t is the time; ω is 2π times the signal frequency; and H_0 is a measure of the maximum magnetic field in the guide. Hence the amount of power transported varies with $|H_0|^2$. In general, H_0 is a complex number which may also be used to specify the phase of the wave when $t = 0$.

β is the *propagation constant* and defines how the phase of the wave varies along the length of the guide. m and n are non-negative integers which are called the *mode numbers*. In fact, in metal waveguides of this type the mode $m = n = 0$ cannot exist as this would require a non-zero electric field along a guide wall.

From the expressions given above, we can deduce some of the general properties of guided waves. It can be seen that there are a range of possible TE waves corresponding to the choice of m and n. These are conventionally distinguished by calling the solution for particular values the TE_{mn} *mode*.

It can be seen that each mode has a unique field distribution in the (x, y) plane and that the form of this distribution does not change as the mode propagates along the guide. It is also important to note that the propagation constant depends upon the mode numbers. As a consequence the effective wavelength along the guide varies from mode to mode.

It is a general property of the solutions of a differential equation that, if A_1, A_2, A_3, etc, are solutions, then any linear combination of them is also a solution. Hence the wave in a guide may consist of various amounts of a number of modes, each moving along with its own effective wavelength and keeping its own characteristic mode field pattern.

For most purposes it is undesirable to have a multi-mode signal. Although each mode propagates keeping its own field distribution, the total field at any point in the guide is produced by adding together the contributions of the various modes present. As the mode propagation constants differ, this means that the actual field distribution at a given plane depends upon where we are placed along the guide.

A signal sensor placed in the guide will have been designed to respond to a particular field pattern. If the transmitted wave does not match the desired pattern, the sensor will not work efficiently and some signal power will be lost. Furthermore, the propagation constants of the modes are all frequency dependent. Hence the signal losses will display a frequency sensitivity which make the sensor difficult to use over a wide frequency range.

In principle it is possible to design a sensor which matches the multi-mode field in some specific circumstances, but in general it is preferable to produce a system where only one mode can be transmitted. This makes design much simpler and usually means that better performance can be obtained with much less effort.

For a rectangular metallic waveguide we find that the propagation constant is zero when the signal frequency is such that

$$\omega^2 \mu_0 \varepsilon_0 = k_c^2. \tag{1.6}$$

At frequencies below this value the propagation constant is imaginary. This means that the field in the guide decays exponentially with z rather than being a sinusoid, i.e. the wave cannot propagate along the guide. The mode is said to be *cut off* at this frequency.

The frequency at which β becomes zero depends upon the guide width and height, and also upon the mode numbers. Microwave engineers have exploited this behaviour to develop what is referred to as a *standard rectangular waveguide*. This has a standard width-to-height ratio of 2 to 1.

We can characterize the cut-off point of a mode in terms of the wavelength in free space which corresponds to the frequency at which the propagation constant becomes zero. If we choose a guide where $a = 2b$ the cut-off wavelengths λ_c of the first few modes will be

$$\begin{aligned}
&m = 1 \quad n = 0 \quad &\lambda_c = 2a \\
&m = 0 \quad n = 1 \quad &\lambda_c = a \\
&m = 1 \quad n = 1 \quad &\lambda_c = (0.89...)a \\
&\text{etc.}
\end{aligned}$$

As the mode numbers increase, so the cut-off wavelength reduces. In the wavelength range $2a > \lambda > a$, only *one* mode is possible. Over the corresponding range of frequencies the guide is referred to as being *single mode*. In this range the field shape and propagation rate can be uniquely defined within the guide. This greatly simplifies the design of waveguide-based systems.

A standard rectangular waveguide used as a single-mode transmission system is widely used at microwave frequencies. However, it requires the guide width and height to be around one half to one quarter of the free-space wavelength of the radiation. This becomes increasingly difficult to achieve at millimetre-wave frequencies as it requires us to manufacture rectangular pipes with dimensions of a millimetre or less.

Even when we are able to make such small guides, we discover that their performance is poor compared with the larger guides used at lower frequencies. In part this is due to problems in accurately making the guide, but we also encounter a more fundamental problem.

As the wave moves down the guide, it sets up currents in the surface of the walls. These currents must exist if the walls are to reflect the wave and to confine it to the guide. As was mentioned in the Preface, the velocity of an electromagnetic wave is far higher in air than in a normal metal. Only electrons near the surface have time to respond to the field before it is reflected. As the signal frequency rises, the depth of the layer in which the electrons respond becomes less.

In effect we find that the electrons are confined to a thinner conductor layer at higher frequencies. In a normal resistive material this means that the resistance through which the wall current must flow increases with frequency. Similarly, as we try to reduce the guide size and to maintain single-mode operation, we are reducing the effective width of the conductor available for the wall current, increasing the resistance still further.

The losses in normal microwave guides can be almost unmeasurably small, but at frequencies around 100 GHz a standard waveguide may have a loss approaching $1\,\mathrm{dB\,cm^{-1}}$ or more. These losses rise rapidly at higher frequencies, making a standard waveguide unusable except over very short distances.

Optics and laser engineers also need to be able to transmit electromagnetic waves from place to place as efficiently as possible. Instead of using metal waveguides they employ either beams in free space or various forms of dielectric structures, e.g. fibre guides.

Dielectric fibres (and similar structures) can work very well for visible and near-visible frequencies. Unfortunately there are, as yet, no millimetre-wave dielectrics which have come anywhere near matching the low losses of the best materials for the near-visible region. Free space is, by its very nature, a low-loss transmission medium—and it does not present us with any difficulties of supply or manufacture! For this reason it has proved possible to design millimetre-wave systems which work very well over a wide frequency range using signal coupling by beams in free space.

Visible and near-visible radiation is typified by a wavelength of less than a micron. It is usually not very difficult to make systems of optics for visible light where the sizes of the beams, lenses, etc, are many thousands of wavelengths or more. These systems can, for most purposes, be analysed or designed using the traditional methods of ray optics. Light can be treated as being transmitted from place to place via plane-wave beams.

1.2 Gaussian Beams in Free Space

In order to treat millimetre-wave optical systems in the same way, we would have to use beams and optical elements which were perhaps a metre or more across. This would make millimetre-wave optical systems impractical for all but a few purposes. Clearly we need to be able to design *compact* millimetre-wave optical systems—ones where the beam and optical element sizes are no more than a few tens or hundreds of wavelengths.

During the 1960s, laser engineers were coming to grips with designing optical systems which were not many wavelengths across. At much the same time other workers were looking into the possibility of using chains of lenses to couple radiation from place to place. Out of this work grew the subject of *Gaussian beam mode optics* which has proved to be an outstanding method for analysing compact optical systems. (Optical engineers also use a technique called *Gaussian ray optics*. It is important to note that is *not* the same as the beam mode technique which is being discussed here.)

Gaussian beam mode (GBM) theory generally makes the following assumptions:

(i) that the radiation is moving in a paraxial beam whose cross sectional size is *not* sufficiently large that it may be treated as plane parallel;
(ii) that the radiation can be represented as a *scalar* field distribution.

By 'paraxial' we mean that the signal is moving essentially in a beam along a given axis but that some diffraction is taking place. Having represented the field as a scalar we can then associate that scalar magnitude and phase with, say, a specific electric vector component of the actual electromagnetic wave. These assumptions allow us to simplify Maxwell's equations and to obtain a solution appropriate to a beam moving through space. It should be noted, however, that this means the results are only useful in real situations which mimic the assumptions.

We can write the scalar version of the wave equation as

$$\nabla^2 \psi + k^2 \psi = 0 \tag{1.7}$$

where $k = 2\pi f/c$. ψ represents the scalar field appropriate for the beam, c is the velocity of light and f is the signal frequency.

For a beam moving essentially paraxially in the z direction (Cartesian coordinates) we can define a function u such that

$$\psi = u(x, y, z) \exp(-jkz) \exp(j2\pi ft). \tag{1.8}$$

Since the beam is paraxial along the z direction, we can assume that virtually all the z dependence of ψ is contained in the exponential terms which multiply u above. Hence we can assume that

$$\frac{\delta^2 u}{\delta z^2} \ll \frac{\delta^2 u}{\delta x^2} + \frac{\delta^2 u}{\delta y^2}. \tag{1.9}$$

From the scalar wave equation we can therefore write

$$\frac{\delta^2 u}{\delta x^2} + \frac{\delta^2 u}{\delta y^2} - 2kj \frac{\delta u}{\delta z} = 0. \tag{1.10}$$

The solutions of this differential equation are known. From them we can say that

$$\psi_{mn} = \frac{1}{\omega} E \mathcal{H}_m\left(x\frac{\sqrt{2}}{\omega}\right)\mathcal{H}_n\left(y\frac{\sqrt{2}}{\omega}\right)\exp\left[-j(kz - \Phi_{mn} - 2\pi ft) - r^2\left(\frac{1}{\omega^2} + \frac{jk}{2R}\right)\right] \quad (1.11)$$

where $\mathcal{H}_m$ and $\mathcal{H}_n$ are Hermite polynomials of degree m and n of the form defined in appendix 1,

$$\Phi_{mn} = (m+n+1)\arctan\left(\frac{\lambda z}{\pi\omega_0^2}\right) \quad (1.12)$$

$$\omega^2 = \omega_0^2\left[1 + \left(\frac{\lambda z}{\pi\omega_0^2}\right)^2\right] \quad (1.13)$$

$$R = z\left[1 + \left(\frac{\pi\omega_0^2}{\lambda z}\right)^2\right] \quad (1.14)$$

and

$$r^2 = x^2 + y^2. \quad (1.15)$$

E is a complex number which defines the overall amplitude and phase of the beam at $t=0$. The beam power will be proportional to $|E|^2$.

As we found for the case of a metallic waveguide, the solutions represent a set of modes. Each mode propagates keeping its own specific form and the rate of change in phase along the beam depends upon the mode numbers m and n. As before, we may expect that a single-mode beam would be preferred to a multi-mode beam as this ensures that the field profile at any place along the beam is easy to define and is not strongly frequency dependent.

Most practical systems involve beams with circular or square symmetry, i.e. the beam sizes are identical in the x and y directions at any place along the beam. Equations (1.11)–(1.14) are in the form appropriate for a beam of this type. In some cases the beam may have elliptic or rectangular symmetry and has a beam size, phase-front curvature, etc, which differ in the x and y directions. Under these more general circumstances we may define the beam as being the product

$$\Psi(x,y) = \Psi(x)\Psi(y)\exp[-j(kz - 2\pi ft)] \quad (1.16)$$

where $\Psi(x)$ and $\Psi(y)$ are linear combinations of one-dimensional mode expressions of the form

$$\Psi_m(x) = \sqrt{\frac{1}{\omega_x}} E_x \mathcal{H}_m\left(x\frac{\sqrt{2}}{\omega_x}\right)\exp\left\{-j\Phi_m - x^2\left[\left(\frac{1}{\omega_x}\right)^2 + \frac{jk}{2R_x}\right]\right\} \quad (1.17)$$

where the subscript x denotes the value appropriate for the x direction. The

Signal Transmission, Modes and Gaussian Beams

beam size ω_x and phase-front curvature R_x may be obtained from equations (1.13) and (1.14) simply by replacing ω_0 with ω_{x0}. For the one-dimensional case,

$$\Phi_m(x) = (m + \tfrac{1}{2}) \arctan\left(\frac{\lambda z}{\pi \omega_{x0}^2}\right). \qquad (1.18)$$

Although free-space modes share many properties with modes propagating along a metallic waveguide, there are a few important differences.

The $m = n = 0$ mode *is* allowed in free space. Indeed, this mode—usually called the *fundamental* mode—proves to be the best for many purposes. It is fairly easy to produce and detect. It also proves to be the best choice when the effects of truncation by apertures are taken into account.

There is no cut-off wavelength for a free-space mode. The propagation constant—which determines the rate of phase change along the beam axis—cannot be zero unless the signal frequency or beam cross sectional size go to zero. Also unlike the metal waveguide modes, the propagation constant of a free-space mode is *not* constant along the beam. From equation (1.11) we may see that changes in phase along the beam axis of a Gaussian mode may be obtained from

$$\theta = kz - \Psi_{mn} \qquad (1.19)$$

where θ represents the phase at z. The term Φ is sometimes called the *anomalous phase term* as it indicates a phase different between a Gaussian mode and a plane wave.

From appendix 1 the fundamental mode is of the form

$$\psi_0 = \frac{1}{\omega} E \sqrt{\frac{2}{\pi}} \exp\left\{-j(kz - \Phi_0 - 2\pi ft) - r^2\left[\left(\frac{1}{\omega}\right)^2 + \frac{jk}{2R}\right]\right\}. \qquad (1.20)$$

(Note that we have adopted the convention of using a single subscript zero to represent the fundamental-mode beam.)

The expressions used above have been simplified by assuming that the beam axis is along the line $x = y = 0$. Hence ψ is the field at a place a distance r from the beam axis. For the fundamental mode we can see that the field amplitude varies as

$$\exp\left(-\frac{r^2}{\omega^2}\right) \qquad (1.21)$$

i.e. the amplitude variation across the beam has a Gaussian shape. ω is a measure of the width of the distribution at any place along the beam. For this reason, ω is often referred to as the beam *size* or the beam *radius*.

In the special case of the fundamental mode, the beam radius ω is also the distance from the beam axis at which the field amplitude has fallen to $1/e$ of its value on the beam axis.

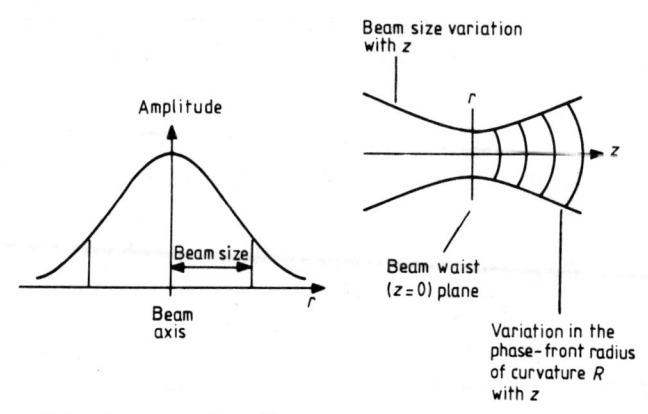

Figure 1.2 Cross section of Gaussian field pattern and variations of size and phase-front curvature along the beam.

Figure 1.2 shows how the field of a fundamental-mode Gaussian beam varies across the beam. It also illustrates how the beam radius ω varies along the beam. As this variation is hyperbolic, ω has a minimum value at some plane along the beam. This plane is called the *beam waist plane* and the value of the beam size at this plane is defined as ω_0, the *beam waist radius* (or size). For simplicity, all the expressions given above are for where coordinate origin is placed such that $z = 0$ is the beam waist plane.

The beam is not plane parallel. Its amplitude distribution is not uniform across the beam and its size varies across the beam. Hence its phase front cannot, in general, be plane. The phase distribution in a given plane depends upon the parameter R in the above expressions. The form of this plane variation is that of a spherical phase front of radius R.

Although the width of the field distribution varies along the beam, the shape remains Gaussian. Similarly, although the phase-front curvature alters, it remains spherical. For higher-order (i.e. $m \neq 0$ or $n \neq 0$) modes the shape may be different but the same rules apply, i.e. the beam shape does not alter along the beam.

Throughout most of this book it will be assumed that the beams being considered are composed of just a single (usually the fundamental) mode with circular or square symmetry. Most of the arguments and methods discussed can, however, be applied to a multi-mode beam. The advantages of a single-mode beam are that the computation is generally easier and the overall performance of a practical system is simpler. The details of multi-mode analysis are considered in chapter 3.

Chapter 2

Beam Coupling, Lenses and Mirrors

2.1 Coupling between Gaussian Beams

In the previous chapter we derived the basic expressions which define the way in which a beam of electromagnetic power propagates in free space. In this chapter we see how we can use beams to couple this power efficiently from one component to another of a millimetre-wave system.

Many millimetre-wave devices are mounted in a small length of standard rectangular waveguide. A typical system will consist of sources, detectors, etc, each in a short piece of waveguide. The advantage of this method is that it allows us to define the field pattern surrounding the device, making design and analysis easier. In order to avoid the problems mentioned in chapter 1, however, the signals being coupled between components should be carried by free-space beams over as much of the intervening distance as possible.

The device's waveguide must be terminated by an arrangement designed to couple the field propagating along the guide to a free-space beam. We require a suitable *antenna* for the devices. We also must be able to couple the beam produced by a source antenna into that required by another antenna attached to a detector in order to communicate signal power between them. The subject of antennas will be discussed in chapter 4. Here we shall look at how we can couple power between beams in free space.

Figure 2.1 illustrates the coupling of two beams in space. Clearly, we may expect that any misalignment of the two beams will limit the efficiency of signal power coupling. Here we shall therefore make the simplest possible assumption, namely that the beams share a common axis of propagation. The two beams can then differ in the following ways.

(i) Their beam waist sizes may differ.
(ii) Their beam waist planes may not coincide.
(iii) The beam profiles may not be the same (i.e. they may be different modes or be composed of different linear combinations of modes).

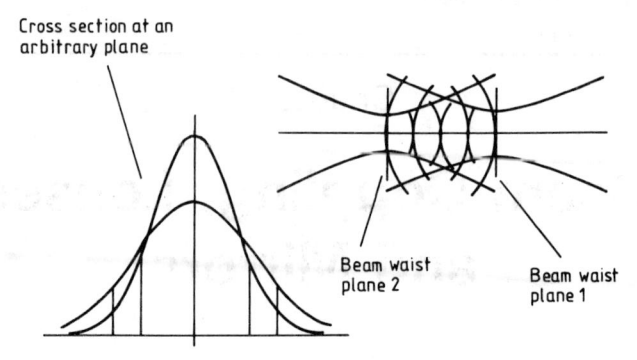

Figure 2.1 Field patterns of two beams with the same axis but different waist positions.

In order to obtain perfect coupling, the two beams should be identical. Then, power in one beam is indistinguishable from power in the other, and is transferred with ideal efficiency. In practice, however, the beams may differ in one or more of the respects mentioned above.

Consider the case where, at an arbitrary plane normal to the beam axes, the two field distributions are Φ and Ψ respectively. Each of these is a Gaussian mode (or combination of modes). In this example we may take Φ to represent the beam field pattern produced by a signal power source and Ψ to be the pattern of sensitivity to field of a detector.

At a point (x, y) on our plane, $\Phi(x, y)$ represents the field produced by the source. $\Psi(x, y)$ represents the sensitivity of the detector to field at that point. Both Φ and Ψ are complex numbers which define the magnitude and phase distribution of the fields.

What fraction of the power contained by the beam of field Φ will be coupled into the detector beam of field Ψ and hence into the detector? Since the fields are complex, we determine the power coupling from calculating the total field coupling between the beams. This can be done by integrating the incremental amount of coupling for each small area (dx, dy) over the plane where the fields are known.

The total field coupling will be

$$\int_{-\infty}^{+\infty} \int_{-\infty}^{+\infty} \Phi \Psi^* \, dx \, dy. \tag{2.1}$$

It should be noted that the integral is in terms of the product of one field and the complex conjugate of the other.

The total power coupled into the detector beam will therefore be

$$P = \left| \int_{-\infty}^{+\infty} \int_{-\infty}^{+\infty} \Phi \Psi^* \, dx \, dy \right|^2. \tag{2.2}$$

The power contained by the input beam can be defined to be the integral of the modulus squared of its field distribution, i.e. the source power may be defined as

$$S = \int_{-\infty}^{+\infty} \int_{-\infty}^{+\infty} \Phi \Phi^* \, dx \, dy. \tag{2.3}$$

The *power-coupling efficiency* N can then be defined to be the ratio

$$N = \frac{P}{S}. \tag{2.4}$$

We would associate a power-coupling efficiency N of unity with perfect loss-free coupling. From this definition and the expressions given above, we require that

$$\int_{-\infty}^{+\infty} \int_{-\infty}^{+\infty} \Psi \Psi^* \, dx \, dy = 1 \tag{2.5}$$

i.e. for the above expressions to be true the detector beam pattern must be normalized to the amplitude of a beam which has unity overall power.

The actual coupling efficiency between the source and detector beams in a specific practical system cannot depend upon which plane we chose for calculating the coupling integrals. In reality a definite amount of power is coupled and our calculation must yield this result if it is to prove correct. Hence we may expect that the choice of coupling plane should not alter the value of N that we obtain. Similarly, our choice of coordinate system should not alter the result.

It is convenient when considering the coupling integrals which must be evaluated to obtain a power-coupling efficiency to use an expression which avoids being specific about the coordinate system. This can be used to make the actual expression more compact and the argument clearer. It is also a useful continuing reminder that the choice of coordinate system can be made on the basis of computational convenience in a particular case. Here we shall therefore adopt the notation that, in Cartesian coordinates,

$$\langle \Phi | \Psi \rangle = \int_{-\infty}^{+\infty} \int_{-\infty}^{+\infty} \Phi \Psi^* \, dx \, dy. \tag{2.6}$$

$\langle \Phi | \Psi \rangle$ can now be used to represent the integral product over the plane irrespective of the coordinate system.

We may now redefine the power-coupling efficiency N via the expression

$$N = \frac{|\langle \Phi | \Psi \rangle|^2}{\langle \Phi | \Phi \rangle \langle \Psi | \Psi \rangle}. \tag{2.7}$$

Note that equation (2.7) also takes into account the possibility that the field distribution Ψ has not already been normalized in accord with equation (2.5).

Having obtained a suitable definition for power-coupling efficiency, we may use it to determine the fraction of power coupled between beams in specific cases.

As an example, let us take the simplest case, that of the coupling between two fundamental Gaussian modes. In Cartesian coordinates we may write

$$\Phi = \frac{1}{\omega} \exp\left\{-r^2\left[\left(\frac{1}{\omega}\right)^2 + \frac{jk}{2R}\right]\right\} \quad (2.8)$$

$$\Psi = \frac{1}{\omega'} \exp\left\{-r^2\left[\left(\frac{1}{\omega'}\right)^2 + \frac{jk}{2R'}\right]\right\}. \quad (2.9)$$

Since we are concerned here only with the power-coupling efficiency, we can omit those terms which only affect the phase of the coupling integral. In these expressions we have used ω and R to represent the beam size and phase-front curvature of one beam and ω' and R' the corresponding parameters for the other beam.

We may now say that

$$\langle \Phi | \Psi \rangle = AB \int_{-\infty}^{+\infty}\int_{-\infty}^{+\infty} \exp\left\{-r^2\left[\left(\frac{1}{\omega}\right)^2 - \left(\frac{1}{\omega'}\right)^2 + \frac{jk}{2}\left(\frac{1}{R} - \frac{1}{R'}\right)\right]\right\} dx\,dy \quad (2.10)$$

$$\langle \Phi | \Phi \rangle = A \int_{-\infty}^{+\infty}\int_{-\infty}^{+\infty} \exp\left[-2\left(\frac{r}{\omega}\right)^2\right] dx\,dy \quad (2.11)$$

$$\langle \Psi | \Psi \rangle = B \int_{-\infty}^{+\infty}\int_{-\infty}^{+\infty} \exp\left[-2\left(\frac{r}{\omega'}\right)^2\right] dx\,dy \quad (2.12)$$

where

$$A = \frac{1}{\omega} \quad (2.13)$$

and

$$B = \frac{1}{\omega'}. \quad (2.14)$$

By consulting a suitable table of standard integrals we find that

$$\int_{-\infty}^{+\infty} \exp(-at^2)\,dt = \sqrt{\frac{\pi}{a}} \quad (2.15)$$

from which we can obtain the result

$$N = \frac{4}{(\omega/\omega' + \omega'/\omega)^2 + (k\omega\omega')^2(1/R' - 1/R)^2}. \quad (2.16)$$

As we would expect, equation (2.16) confirms that N will only be unity when $\omega = \omega'$ and $R = R'$, i.e. when the two beams are identical. The

Beam Coupling, Lenses and Mirrors

method employed to obtain this result may be applied for beams composed of other single- or multi-mode beams (see, for example, chapter 3).

In most practical cases we are presented with two beams which are essentially fundamental-mode Gaussians but which may differ in beam waist size and location. Provided that the beams have a common axis there will always be at least one plane where their sizes are equal. It is at this plane that we may choose to place a lens or mirror in order to optimize the beam coupling.

2.2 Lenses and Beam Coupling

A lens may be regarded as a device which alters the radius of phase-front curvature of a beam passing through it. A suitable lens for our purposes changes the radius of curvature of an incident beam so that it matches exactly that of another beam into which we wish to couple power. The modified input beam and the output beam now have the same sizes and curvatures and—barring other problems—the coupling efficiency would be unity.

For a 'thin' lens the focal length f may be defined simply in terms of the required change in phase-front curvature:

$$\frac{1}{f} = \frac{1}{R} - \frac{1}{R'}. \qquad (2.17)$$

(It is assumed that the sign of R determines on which side of the lens it is centred.)

In practice, all lenses possess various defects which may affect the performance of the system in which they are placed. These may be summarized as follows:

(i) distortions and aberrations;
(ii) absorption losses;
(iii) surface reflections.

Equation (2.16) gives the coupling efficiency between two fundamental mode beams which may have differing sizes and curvatures. If the lens which seeks to equalize the curvatures alters the field pattern so that it is no longer simply Gaussian, then equation (2.16) ceases to be the appropriate expression and the coupling efficiency will be reduced.

In conventional ray optics we could draw a fan of rays from a focal point to the surface of a lens. The directions of the locally refracted rays may then be obtained from Snell's law and the focusing action—and distortions—of the lens obtained. By applying this method, appropriate lens profiles can then be calculated.

For a Gaussian mode we may assume that power propagates locally in the direction normal to the beam phase front. As the phase front is everywhere spherical, this means that at any plane normal to the beam axis the power appears to diverge or converge as if from a point source. However, the apparent location of this point *varies* as we move along the beam. This would not be the case for a 'ray optics' beam and means that a lens designed using simple ray optics may not work as expected when the beam is compact.

From equation (1.14) we may see that, when $z \gg \pi\omega_0^2/\lambda$, then $R \simeq z$ and the phase front appears essentially fixed at the centre of the beam waist plane. This condition is equivalent to saying that the beam size is much larger than the wavelength and the normal assumptions of ray optics can be made. In general, however, we must take the Gaussian nature of the beam into consideration when designing lenses if we wish to avoid unexpected distortions and coupling losses.

Figure 2.2 illustrates a Gaussian mode incident upon a lens surface. On the beam axis the change in phase between the beam waist plane and z_p is

$$\alpha = -kz_0 + \arctan\left(\frac{\lambda z_0}{\pi\omega_0^2}\right) - k\mu(z_p - z_0) \tag{2.18}$$

and at the point where z_p intersects the lens surface the phase change compared with the waist plane is

$$\beta = -kz_p + \arctan\left(\frac{\lambda z_p}{\lambda\omega_0^2}\right) - \frac{kr^2}{2R}. \tag{2.19}$$

It is assumed that the lens is in air and that μ is the refractive index of the lens material. It has also been assumed that the beam is in the fundamental mode. One of the useful properties of Gaussian modes is that R and ω do not depend upon the mode numbers. However, the anomalous phase term $\bar{\Phi}$ *does* depend upon the mode numbers and this may need to be taken into account in some cases.

If we arrange that these phase changes will be equal for every value of the distance r from the beam axis, then we have essentially produced a plane wave within the lens. From the above expressions it can be seen that the requirement for us to obtain this result is that

$$r^2 = \omega_0^2(\hat{z}_p + \hat{z}_p^{-1})(\Gamma + \arctan\hat{z}_p - \arctan\hat{z}_0) \tag{2.20}$$

where

$$\hat{z}_p = \frac{2z_p}{k\omega_0^2} \tag{2.21}$$

$$\hat{z}_0 = \frac{2z_0}{k\omega_0^2} \tag{2.22}$$

Beam Coupling, Lenses and Mirrors

and
$$\Gamma = \tfrac{1}{2}(\mu - 1) k^2 \omega_0^2 (\hat{z}_p - \hat{z}_0) \qquad (2.23)$$

These expressions link r with $z_p - z_0$ and allow us to calculate a lens surface which produces a nominally plane beam inside the lens. A real lens, curved on both faces, can be treated as a 'back-to-back' combination of a pair of such lenses. Each produces the desired free-space beam and the two lens surfaces are coupled via an essentially parallel beam within the lens.

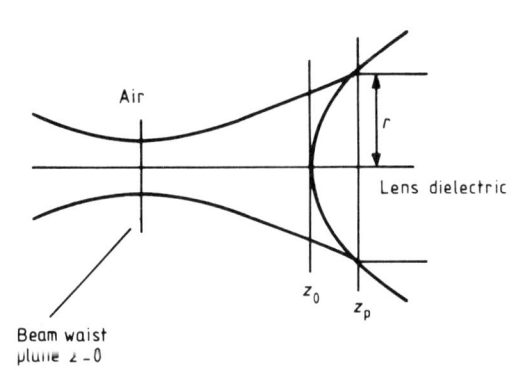

Figure 2.2 Focusing effect of a convex lens surface.

A lens of this type will work well provided that it is 'thin', i.e. provided that its axial thickness is small compared with its effective focal length. When this is not the case, problems will arise which distort the beam and reduce the reliability of the calculation.

(i) Some of the power incident upon a dielectric interface will be reflected. The amount reflected depends upon the angle of incidence. On the axis the power arrives at normal incidence. As we move away from the axis the angle of incidence steadily changes, altering the reflectivity. This tends to increase the total loss and distorts the amplitude pattern of the transmitted beam.

(ii) It is inevitable that the lens material will absorb some of the power passing through it. For example, in a convex lens this means a higher absorption loss near the axis than at the edges.

(iii) Power falling on the surface at an angle appears 'foreshortened'. The curved geometry of the refractive surface will alter the field distribution of the beam passing through it.

(iv) We may no longer regard the beam within the lens as being essentially parallel. Instead we must treat it as a Gaussian beam and take diffraction effects into account. This problem is, however, rarely significant unless the lens is exceptionally thick and has a small diameter compared with the radiation wavelength.

2.3 Dielectric Materials and Surface Reflections

Provided that the lens thickness-to-focal length and diameter-to-focal length ratios are less than about 0.2 the geometric losses are less than those due to absorption or reflection for most millimetre-wave lenses. Reflection and absorption losses depend upon the lens material and the signal frequency.

Table 2.1 summarizes the dielectric properties of some of the lower-loss millimetre-wave dielectric materials. It is convenient to quote the loss tangent of the materials since, in general, they have an absorptivity which is proportional to the signal frequency over the millimetre-wave region. At a frequency f in gigahertz the loss α of a material in decibels per centimetre may be obtained from the loss tangent δ via the expression

$$\alpha = 0.91 \mu \delta f$$

where μ is the refractive index.

The values quoted for quartz and sapphire in table 2.1 were obtained from crystalline samples. Two values are given as these crystals are birefringent. This makes them unsuitable for lenses in some cases as the lens properties will be polarization sensitive. The materials may be useful, however, in cases where a particularly high refractive index is required.

High-density polyethylene (HDPE) and polytetrafluoroethylene (PTFE) have about the lowest losses known for millimetre-wave dielectrics. Although they cannot be polished or ground like quartz or sapphire, they can be turned quite easily on a lathe. For mass production purposes they could also be pre-cast to the required shape. Both materials are reasonably strong and inert.

TPX (poly 4 methyl pentane-1) is another polymer material which has been used on a number of occasions to manufacture millimetre-wave lenses. Although significantly more lossy than either HDPE or PTFE (by a factor of 10) it does possess one useful property. TPX is reasonably transparent

Table 2.1 Dielectric properties of some millimetre-wave materials.

Material†	Refractive index	Loss tangent ($\times 10^3$)
HDPE	1.53	0.3
LDPE	1.51	0.3
PTFE	1.43	0.8
Quartz	2.11, 2.15	0.3, 0.8
TPX	1.44	4.0
Sapphire	3.06, 3.42	1.7, 2.9

† The abbreviations are given in full in the text.

to visible light and has a refractive index in the visible similar to its millimetre-wave index. For this reason, systems including TPX lenses may be aligned by eye or using visible lasers as 'ray' sources. HDPE and PTFE are opaque white materials and their use prevents alignment in this way.

Low-density polyethylene (LDPE) has millimetre-wave properties which are virtually identical with those of HDPE. It is, however, a relatively soft material, which is difficult to machine.

There is evidence that the millimetre-wave dielectric properties of most commercial polymers are somewhat variable. Manufacturers are usually more concerned to produce a material suitable for milk crates than for millimetre-wave lenses! Most commercial polymers will include small amounts of other materials intended to improve, for example, the way in which the material flows when being pressure moulded. The polymerization processes followed at different times by various manufacturers are also varied. This can produce 5–10% variations in refractive index—and may cause the absorptivity to rise in some cases by an order of magnitude. In most cases the dielectric constants are close to the values quoted in table 2.1, but it is a good idea to select lens materials with caution for demanding applications.

The reflectivity of a dielectric interface depends upon the refractive indices and the angle of incidence. A wave moving through a medium of refractive index μ_1 and striking another medium of refractive index μ_2 at normal incidence will produce a reflected field whose magnitude is

$$E_r = \frac{\mu_2 - \mu_1}{\mu_2 + \mu_1} \times E_i \qquad (2.24)$$

times the incident field magnitude.

For example for an air–HDPE interface this means that we may expect a 0.23 dB loss in the power transmitted due to reflection. This compares with an absorption loss in the material of around 0.1 dB cm^{-1} at 250 GHz. For a dielectric lens in air a reflection loss will occur at both of the lens surfaces. The reflected waves will interfere and the total reflection loss then depends upon the lens size and shape compared with the radiation wavelength.

The effects of lens reflections will be considered in more detail in a later chapter. Here we may simply look at the general behaviour by regarding a thin lens as being essentially a plane parallel-sided slab of dielectric.

The power reflection coefficient $|\Gamma|^2$ of such a slab will be

$$|\Gamma|^2 = \frac{(r_1 + r_2)^2 - 4r_1 r_2 \sin^2 x}{(1 + r_1 r_2)^2 - 4r_1 r_2 \sin^2 x} \qquad (2.25)$$

where r_1 and r_2 are the field reflectivities of the two surfaces, and

$$x = \frac{2\pi d}{\lambda} \sqrt{\mu^2 - \sin^2 \theta}. \qquad (2.26)$$

d is the slab thickness, μ its refractive index, λ the radiation wavelength in free space and θ the angle of incidence of the beam moving inside the slab. For simplicity the beam has been assumed to behave as if plane parallel and the slab absorption loss is neglected.

For such a slab,

$$r_1 = -r_2. \qquad (2.27)$$

At normal incidence the reflectivity varies with d/λ between a maximum value

$$|\Gamma|^2(\text{max}) = \frac{4r^2}{(1+r^2)^2} \qquad (2.28)$$

(where r can be taken as being equal to either r_1 or r_2), and a minimum value of zero. For HDPE, $r = 0.21$, and $|\Gamma|^2(\text{max}) = 0.16$, leading to a maximum power loss by reflection of around 0.75 dB.

From this example it may be seen that reflection losses can be significantly higher than those produced by absorption. They are also strongly frequency dependent because of the interference effects between the waves reflected from different surfaces.

Various methods have been evolved to reduce the reflection losses of millimetre-wave lenses. The simplest is to choose carefully a lens thickness which causes the two surface reflections to cancel as nearly as possible. Whilst this method works fairly well, it suffers from two drawbacks. Firstly, the surfaces of a lens are *not* plane parallel. The reflected waves cannot be made to overlay exactly and to cancel perfectly. This means that it may not be possible to obtain a true zero total reflection, although the reflections may be significantly reduced. Secondly, a typical lens will have a thickness which is somewhat longer than a wavelength. Adjacent maximum and minimum values of $|\Gamma|^2$ are separated by a change of just 0.25 in the ratio $d\mu/\lambda$. Consider the example of an HDPE lens which is approximately 10 mm thick. This corresponds to around 15 wavelengths at 300 GHz. If the lens is designed to have a minimum reflectivity at 300 GHz, then its power reflectivity will have maxima at around 295 and 305 GHz. Hence reductions in lens reflection produced in this way tend to be useful only when we are dealing with fairly narrow frequency ranges.

Alternative methods are based upon *blooming* (figure 2.3(a)) or *blazing* (figure 2.3(b)) the lens surfaces. This involves placing a suitable layer on each lens surface or patterning the surface so as to reduce the reflection losses. If we place an intermediate dielectric layer, of refractive index μ', in between the air and the lens dielectric, we produce *two* reflective interfaces at the lens surface. The amplitude reflectivities at these two interfaces will now be

$$r_1 = \frac{\mu' - 1}{\mu' + 1} \qquad (2.29)$$

$$r_2 = \frac{\mu - \mu'}{\mu + \mu'}. \qquad (2.30)$$

From equation (2.25) it may be seen that, for normal incidence, the total power reflection may be zero if

$$\mu' = \sqrt{\mu} \qquad (2.31)$$

and the layer thickness t is such that

$$t = \frac{(2n+1)\lambda}{4\mu'} \qquad (2.32)$$

where n is a non-negative integer.

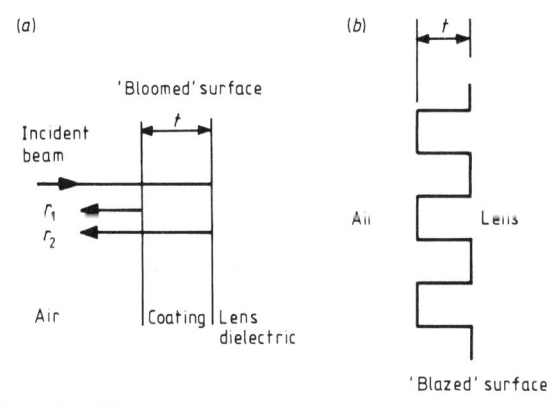

Figure 2.3 Modifications of dielectric surface to reduce reflections.

In this way we may cancel the reflections of each lens surface independently instead of relying upon cancellation effects between different lens surfaces. If the layers are of uniform thickness, the two reflected fields at each lens surface will have essentially identical curvatures, permitting the fields to cancel almost exactly. Furthermore, by choosing $n = 0$ we have selected a layer thickness of just one quarter-wavelength. This means that the reflection reduction extends over a much wider frequency range than would be the case if we seek to employ cancellation effects from surfaces many wavelengths apart. Hence this method is preferable when a lens is to be used over a wide frequency range.

In order to produce successful anti-reflection coatings for HDPE and PTFE lenses we require low-loss materials with a refractive index of around $\sqrt{1.5}$ which can be attached to these materials in uniform layers. In practice this has proved difficult to achieve.

Another method is to cut a series of grooves into each lens surface. Figure 2.3(b) illustrates a series of rectangular grooves cut into a dielectric surface.

The action of this surface contour can be viewed as synthesizing a layer of intermediate refractive index by removing a fraction of the dielectric material. In order to prevent interference from generating scattered waves being reflected or transmitted at an angle to the normal beams, the distance between successive grooves should be small compared with the wavelength.

A typical arrangement would employ grooves at a pitch of $\lambda/4\sqrt{\mu}$, separated by raised areas $\lambda/4(1 + \sqrt{\mu})$ wide, the groove depth being $\lambda/4\sqrt{\mu}$. Concentric grooves (or even a spiral groove) of this type can be cut on HDPE or PTFE using standard lathe machining techniques.

A problem which arises with grooved material is that the effective refractive index does, in fact, depend upon the orientation of the incident wave's electric vector with respect to the groove direction. This may mean that a pattern of grooves will alter the polarization pattern of a beam. Typically blazed layers on lenses are quite thin and hence this effect is not normally significant, but it should be considered where a system is particularly sensitive to unwanted polarization effects.

Modified arrangements have been employed in a number of systems. For example, the polarization sensitivity of grooves may be avoided if the surface contour is produced by drilling an array of holes into the dielectric surface.

Grooves of triangular cross section have also been used. These are particularly useful if they can be cut as a close pattern of deep grooves. The arrangement then behaves as a smoothly graded change in refractive index as the wave moves into the lens material. By cutting grooves (or triangular holes) a few wavelengths deep we avoid any abrupt changes in refractive index and the surface reflectivity may be reduced almost to zero over a very wide frequency range. The distances between adjacent grooves or holes must, however, remain small compared with the wavelength. Unfortunately, it is difficult to manufacture such long narrow grooves or holes in most millimetre-wave dielectric materials.

Two beams may be coupled by placing a lens at the plane where the beam sizes are equal. The action of the lens being to equalize the beam phase-front curvatures, hence maximizing the power coupling. In many cases we are concerned with using a series of lenses (or mirrors) to guide a beam. We must therefore be able to specify how power may be coupled efficiently between focusing elements.

Consider the example of a pair of lenses a distance d apart. The size of the beam emerging from one lens is ω. By altering the focal length of the lens we may alter R, the phase-front radius of the emerging beam, but we cannot alter the beam size at the lens in this way. The beam will have a waist size given by

$$\omega_0^2 = \frac{\omega^2}{1 + \hat{z}^2} \tag{2.33}$$

located a distance

$$z = \frac{R}{1 + (1/\hat{z})^2} \tag{2.34}$$

from the lens, where we have defined

$$\hat{z} \equiv \frac{\pi \omega^2}{\lambda R}. \tag{2.35}$$

The resulting beam at the second lens will have a size ω' and curvature radius R' given by

$$\omega'^2 = \omega_0^2 \left[1 + \left(\frac{\lambda(d-z)}{\pi \omega_0^2} \right)^2 \right] \tag{2.36}$$

$$R' = (d-z) \left[1 + \left(\frac{\pi \omega_0^2}{\lambda(d-z)} \right)^2 \right]. \tag{2.37}$$

These expressions may now be used to select a value for R, the radius of the phase front leaving the first lens, which produces the required beam size ω' at the second.

From equations (2.34) and (2.35) we can see how the beam waist location varies with R. When $R \gg \pi \omega^2 / \lambda$ then $z \simeq R/R^2$ and, as R increases, z falls to zero. When $R \ll \pi \omega^2 / \lambda$ then $z \simeq R$ and, as R decreases, z falls to zero. Rearranging equations (2.34) and (2.35) we obtain

$$R^2 az - R + z = 0 \tag{2.38}$$

where

$$a \equiv \left(\frac{\lambda}{\pi \omega^2} \right)^2. \tag{2.39}$$

Equation (2.38) is a quadratic equation whose roots are

$$R = \frac{1 \pm \sqrt{1 - 4az^2}}{2az}. \tag{2.40}$$

Two conclusions may be drawn from this result. Firstly that the distance z to the beam waist must always be such that

$$z \leqslant \frac{\pi \omega^2}{2\lambda} \tag{2.41}$$

in order to ensure that R is a real number. This means that we cannot produce a beam waist at an arbitrary distance from a lens (or mirror). The distance obtained by setting equation (2.41) to an equality is often called the *maximum throw* of a lens. The second conclusion is that, for a distance to the waist which is less than the maximum throw, we have a choice of *two* solutions for R. These produce different beam waist sizes at the chosen beam waist plane. These are sometimes distinguished by calling the smaller a *focused* beam waist and the larger a *parallel* beam waist.

2.4 Mirrors and Beam Coupling

Metallic mirrors have a number of advantages over lenses. The reflectivity of most metals is essentially unity for millimetre-wave radiation. Mirrors are not subject to the problems of frequency-dependent performance created in lenses by unwanted surface reflections. It is also possible to manufacture very large mirror systems to act as telescope or antenna systems.

The chief disadvantage of the use of mirrors as beam coupling elements stems from their reflection property itself. If used near normal incidence the input and output beams largely occupy the same volume of space. This means that there is a tendency for the source or detector—whichever is the closer—to block a part of the beam coupling the other to the mirror.

Some mirror arrangements, e.g. the Cassegrain system, accept the limitations that this may impose in order to keep the convenience of an axially symmetric optical system. When the mirrors are relatively compact, however, the diffraction problems created by any significant amount of centre blocking are usually better avoided by employing 'off-axis' mirrors.

Mirrors of this type are often considered as if the system was employing just a small part of a much larger reflector well away from its axis of symmetry—hence the term 'off-axis'. The performance of axially symmetric systems such as Cassegrain systems will be considered in the chapter on antennas. Here we shall consider the behaviour of off-axis mirrors used as beam coupling elements.

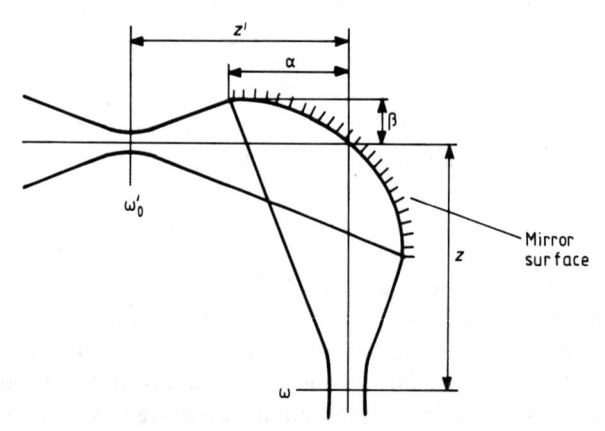

Figure 2.4 Focusing behaviour of an off-axis concave mirror.

Beam Coupling, Lenses and Mirrors

Figure 2.4 illustrates two beams being coupled via a curved reflecting surface. For simplicity we shall examine the case where the input and output beams are at 90° to each other. The mirror surface intersects the beam axis at a distance z from one beam waist and at a distance z' from the other beam waist.

As with the case of a lens we can define the mirror surface by considering the phase variations that we require along the beam. Measured along the beam axis, the phase difference between the beam waist planes will be

$$\delta(\text{axis}) = k(z + z') + \Phi'(z') + \Phi(z) \quad (2.42)$$

and, along the path which touches the mirror surface at a height h above the axial plane of reflection at the point (α, β), the phase shift in moving between waist planes will be

$$\delta(\alpha, \beta, h) = k(z + \beta + z' - \alpha) + \Phi'(z' - \alpha) + \Phi(z + \beta)$$
$$+ \frac{jk\alpha^2}{2R} - \frac{jk\beta^2}{2R'} + \frac{jkh^2}{2R} - \frac{jkh^2}{2R'}. \quad (2.43)$$

The correct mirror surface for coupling the beams whose waists are as shown in figure 2.4 may be obtained by requiring that the phase $\delta(\alpha, \beta, h)$ must equal $\delta(\text{axis})$ for all points on the surface. We may then use the above expressions to determine the mirror profile.

In most cases, α and β are sufficiently small for us to be able to make use of the approximations

$$\Phi(z + \beta) - \Phi(z) \simeq \frac{\delta\Phi}{\delta z}\beta = \frac{\lambda\beta}{\pi\omega_0^2}(1 + \hat{z}^2)^{-1} \quad (2.44)$$

and

$$\Phi'(z' - \alpha) - \Phi'(z') \simeq \frac{\delta\Phi'}{\delta z'}(-\alpha) = -\frac{\lambda\alpha}{\pi\omega_0^2}(1 + \hat{z}'^2)^{-1}. \quad (2.45)$$

In practice, mirrors of this general type tend to produce severe distortions in the amplitude profile of reflected beams due to the off-axis geometry. The shape required as a solution produced by the method described can also be difficult to manufacture. These problems can be eased by arranging that the mirror is sufficiently far away from the beam waists that

$$\frac{\delta\Phi}{\delta z} \simeq 0 \qquad \frac{\delta\Phi'}{dz'} \simeq 0.$$

This enables us to simplify the above expressions which then tend to the form recognizable from conventional ray optics. The mirror surface then becomes part of a standard parabolic, elliptic or hyperbolic curve. However, whilst this helps with design and performance it does so at the expense of accepting a less compact mirror as we have essentially retreated back into the province of ray optics. Failing this we must evaluate the distortions produced in a specific case and decide whether they are acceptable for a given purpose.

The polarization of a beam is also affected by being reflected from a curved off-axis surface. As with the amplitude distortion, this polarization distortion arises from the geometry of the arrangement. Unfortunately, this effect cannot necessarily be reduced if we attempt to retreat to the ray optics domain. The amount of cross polar radiation generated by such a reflector can be a serious problem in a number of applications. It is, for example, one of the reasons why axially symmetric reflecting antennas are preferred for many communications or radar antennas.

In some optical systems it is possible to employ off-axis mirrors in pairs, arranged such that each mirror counteracts the distortions generated by the other. Four possible arrangements are illustrated in figure 2.5. They can be distinguished using two criteria: firstly, whether or not the emergent beam is moving in the same direction as the input; secondly, the choice of a focused or parallel beam waist between the mirrors. Figures 2.5(a) and 2.5(c) show a parallel beam waist between the mirrors, and figures 2.5(b) and 2.5(d) show a focused beam waist. The beam emerges in the same direction that it entered in figures 2.5(c) and 2.5(d) but not in figures 2.5(a) and 2.5(b).

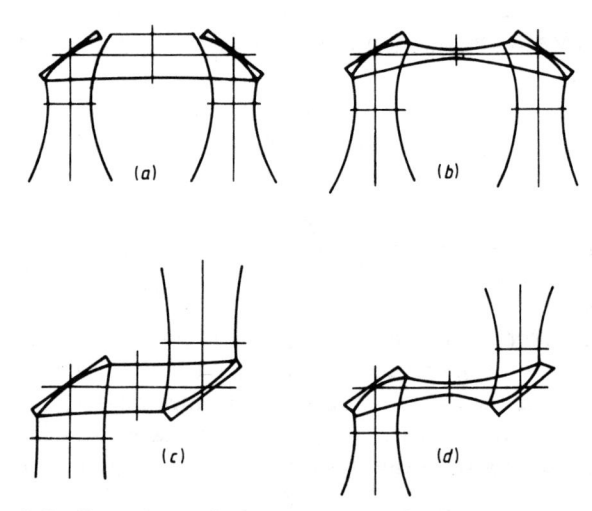

Figure 2.5 Four alternative arrangements for focusing mirror pairs coupled by either a parallel or focused beam waist.

When we wish the output beam to move in the same direction as the input beam, the beam distortions can be minimized by producing a focused beam waist (as shown in figure 2.5(d)) between the mirrors. When the output is moving in the opposite direction to the input beam, a parallel beam waist (as in figure 2.5(*a*)) is to be preferred. These two results may be combined when repeated mirror pairs are used to carry a beam over a long distance. In such a case the waist between each mirror of a pair should be focused and the waist of the beam projected between pairs should be parallel.

Under some conditions it is possible to employ an arrangement of mirrors which are used on axis, without any centre blockage. This allows us to avoid the distortions and losses which usually arise with off-axis reflectors. Figure 2.6 illustrates one type of focusing system which uses axially symmetric mirrors.

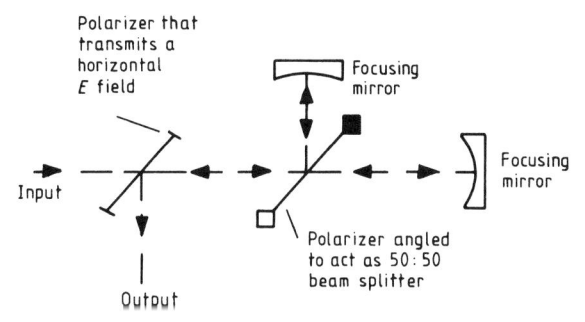

Figure 2.6 Axially symmetric mirror focusing without blockage.

This arrangement uses two wire-grid polarizers and two curved mirrors and is a very simple optical circuit of the sort discussed in more detail in later chapters. In order for this system to operate, the input beam must be plane polarized. (Similar systems may, however, be employed for arbitrary beam polarization states.)

For the system illustrated, the input beam is plane polarized with its electric field parallel to the horizontal plane. Hence it is initially transmitted through the 'horizontal' polarizer. The beam is then incident upon another polarizer which is positioned so as to act as a 50:50 beam splitter. The input beam is hence divided into two orthogonally polarized components of equal amplitudes. These components are reflected by two axially symmetric focusing mirrors and returned to the polarizer which recombines them.

If the path lengths to the two mirrors are identical, the system acts just as if the polarizers were absent and only one mirror were used, i.e. the focused beam is directed back towards the inputs. If we arrange for the path lengths to differ, however, then the polarization state of the reflected beam will depend upon the path difference and the signal wavelength. The system may be considered as a simple form of a polarizing two-beam interferometer, similar to the instruments considered in later chapters.

By choosing a path difference which is $(m + \frac{1}{2})$ wavelengths, where m is an integer, we can arrange that the recombined reflected beam will be plane polarized orthogonally to the input. In these circumstances the output beam is reflected by the horizontal polarizer and is *not* returned to the input. Hence for appropriate signal frequencies the system will focus a beam passing between the input and output.

A system of this form can provide a very low level of beam distortion and loss and this makes it valuable in a number of applications. It has one significant disadvantage: the system is 'tuned' by the choice of a specific path difference. For signal wavelengths other than those which satisfy the above requirement, some power will be directed back towards the input. This particular system is hence unsuitable for signals which cover a wide frequency range.

Chapter 3

Multi-mode Beams and Apertures

3.1 Gaussian Beams and Diffraction

A common problem in the propagation of electromagnetic waves is illustrated in figure 3.1. Here we know the field pattern Ψ on a given plane and wish to calculate the field Ψ' that this produces at some other plane. In conventional ray optics a problem of this type will typically be dealt with using methods such as *Fraunhofer* or *Fresnel* diffraction theory, or by applying a technique known as the *geometric theory of diffraction*. These techniques are valuable, but they do suffer from some practical difficulties.

Consider the situation shown in figure 3.1. We may define the field at a position (x, y) on the first plane to be $\Psi(x, y)$. An incremental area $\mathrm{d}A$ centred on (x, y) will couple a given amount of field onto an incremental area $\mathrm{d}A'$ centred at (x', y') on the second plane, i.e. we may say that, for each choice of (x, y) and (x', y') the field contribution will be such that

$$\delta\Psi'(x', y')\,\mathrm{d}A' = f\{\Psi(x, y)\}\,\mathrm{d}A \qquad (3.1)$$

where the form of the function $f\{\}$ may be derived from the properties of the diffraction theory chosen.

The total field produced at (x', y') can then be obtained by integrating over all the contributions of the various incremental areas in the first plane. Hence we may say that the total field $\Psi'(x', y')$ may be given by

$$\Psi'(x', y') = \int_{-\infty}^{+\infty}\int_{-\infty}^{+\infty} f\{\Psi(x, y)\}\,\mathrm{d}x\,\mathrm{d}y. \qquad (3.2)$$

A double integral is therefore necessary in order to evaluate the field at each point on the second plane.

In many cases of practical interest it is found that the integral does not have a known analytic solution. Then we must resort to a numerical summation over the set of finite incremental areas $\mathrm{d}A$ to calculate Ψ' at a particular place. This means that we have to carry out a four-dimensional summation to calculate the field pattern on the second plane.

With the advent of powerful computers, numerical computations of this type have become increasingly common, being particularly widespread in the areas of antenna analysis and design.

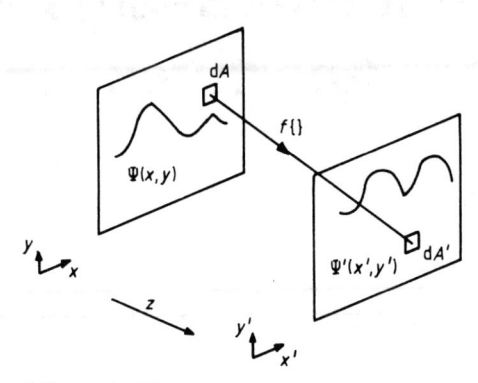

Figure 3.1 Effect of diffraction upon a field pattern moving from one plane to another. z is the nominal direction of propagation for a beam.

When using a method of replacing an integral with summation over a set of finite elements it is necessary to pay some attention to the need to establish the accuracy of the result. Clearly, if our elemental areas are too large and too few, the answers obtained may be unacceptably inaccurate. On the contrary, choosing to employ too many incremental areas produces an unnecessary increase in computational difficulties. Some time must, therefore, be devoted to assessing the optimum choices for obtaining results which are sufficiently accurate. Additional effort may also have to be devoted to checking that the level of accuracy obtained is that which was expected and required. These peripheral requirements can often represent a significant portion of the total effort required to obtain a result.

The numerical approach may also prove unhelpful when the result calculated is not quite what we were looking for. In some cases we wish to obtain an output field Ψ' which differs in some ways from the result obtained. The question which then arises is how we should we alter Ψ to obtain the required result.

When dealing with the propagation of a beam we may wish to know the field patterns produced at a number of planes along the beam direction. We may also wish to find a plane where the field has a particular pattern. This would require us to evaluate repeatedly a four-dimensional summation, searching for the plane that we require. The method may therefore involve us with a considerable amount of computation in order to discover the answer to a fairly simple question.

Multi-mode Beams and Apertures

Where we are interested in power propagating as a paraxial beam we may regard the field Ψ as being produced by a specific combination of Gaussian beam modes. The corresponding fields at any other points along the beam may then be calculated from the properties of Gaussian modes propagating in free space.

This technique possesses two strong advantages over the conventional numerical techniques. Firstly, it allows us to obtain an *algebraic* expression which defines the field magnitude and phase at *any* point on *any* plane along the beam. Once the initial calculation has been performed, we do not need to carry out any further integrals or summations to evaluate the field at some new plane. Secondly, the propagation behaviour of the beam can be described in a way simple enough to allow some physical insight. For example, the manner in which ω and R vary along the beam allow us to identify the optimum position for a coupling lens or mirror rapidly. It also becomes possible in many cases to examine a result and swiftly deduce the effects of various changes.

Consider the field pattern Ψ shown in figure 3.1. We may regard this as being the field produced in a particular plane by a multi-mode Gaussian beam propagating along an axis perpendicular to the plane. The field distribution at this plane is then simply the result of superimposing various amount of different Gaussian modes, i.e. we may write that

$$\Psi = \sum_m \sum_n A_{mn} \psi_{mn} \qquad (3.3)$$

where ψ_{mn} is the mnth Gaussian mode, normalized such that

$$\langle \psi_{mn} | \psi_{mn} \rangle = 1 \qquad (3.4)$$

and A_{mn} is a complex number which determines the amplitude and phase of the contribution of ψ_{mn}.

The modes are all chosen such that, by definition, they all have the same values of beam size ω and phase-front radius of curvature R at the plane where we have defined the field Ψ. This being the case it can be shown to follow that the functions ψ_{mn} are mutually orthogonal, i.e. we may say that

$$\langle \psi_{mn} | \psi_{pq} \rangle = 0 \quad \text{if } m \neq p \text{ or } n \neq q. \qquad (3.5)$$

Now

$$\langle \Psi | \psi_{mn} \rangle = \sum_m \sum_n \langle A_{pq} \psi_{pq} | \psi_{mn} \rangle. \qquad (3.6)$$

So, if we combine equations (3.4)–(3.6), we can obtain

$$A_{mn} = \langle \Psi | \psi_{mn} \rangle \qquad (3.7)$$

which provides us with an expression for calculating A_{mn} once ω and R are chosen.

From ω, R and the signal frequency we may calculate the distance z_0 to the beam waist and the beam waist radius ω_0. Each mode propagates maintaining its own characteristic field pattern. The variations in ω and R as we move along the beam do not depend upon the mode numbers, and all the modes have been chosen to share common values of ω and R at the plane where Ψ is known. Hence all the modes have identical ω and R values at any plane along the beam.

For the mnth mode, the phase change in moving along the beam axis from a plane (ω, R) to the beam waist plane will be

$$\delta(m, n) = kz_0 - \Phi_{mn}(z_0) \qquad (3.8)$$

where $\Phi_{mn}(z)$ is the anomalous phase term for the mnth mode evaluated at the plane a distance z_0 from the beam waist. We can now write an expression of the form

$$\Psi(x, y, z) = \sum_m \sum_n A_{mn} \psi_{mn} \exp(-j\delta) \qquad (3.9)$$

where ψ_{mn} is a Gaussian mode of the form defined in appendix 1. Once we have obtained ω_0, z_0 and the appropriate values of A_{mn}, equation (3.9) is a known algebraic formula which defines the amplitude and phase of the field at *any* point (x, y, z) along the beam.

At first glance it may seem as if the above result implies that the field pattern will change only in size and phase-front curvature as we move along the beam—after all, each mode propagates maintaining its own pattern. R and ω vary along the beam, but their variations are the same for all modes. Such a result would be contrary to our expectation that, except for a few special cases, diffraction will alter the field profile of a beam as it propagates.

Although the field profile of a single-mode beam remains constant along the beam, this is *not* the case for a multi-mode beam. This is because of the presence of the anomalous phase terms Φ_{mn} which *do* depend upon the mode numbers. The total field at any point is obtained by adding together the various mode contributions taking their relative phases into account. The phase relationship will change along the beam, altering the resultant total field pattern.

Figure 3.2 illustrates the alteration in field profile along the beam for a beam composed of a few modes of comparable power levels. For these graphs, $\hat{z}$ is given its standard definition

$$\hat{z} = \frac{\lambda z}{\pi \omega_0^2}. \qquad (3.10)$$

The graphs plotted in figure 3.2(*a*) show the field patterns of four

Multi-mode Beams and Apertures

Gauss–Hermite modes. The graphs in figure 3.2(b) show the field pattern of a multi-mode beam at four planes. The full curves show variation across the beam of the modulus of the field.

The broken curves in figure 3.2(b) show how the phase distribution in a plane may depart from following a spherical phase front. It can be seen that the phase distribution need *not* be simply spherical. This is also a consequence of the changing mode phase relationship as we move along the beam.

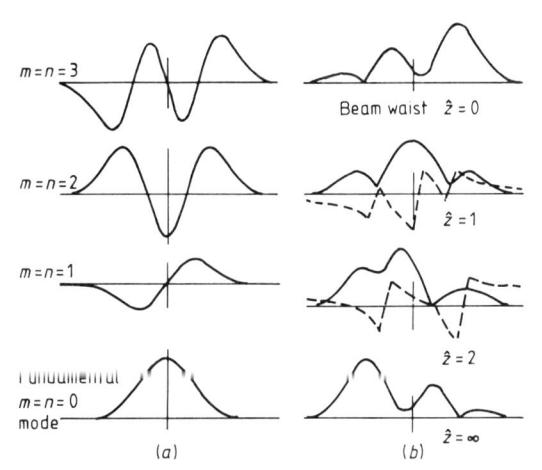

Figure 3.2 (a) The cross sectional field patterns of the first four Gauss–Hermite modes. (b) The total field produced by a beam composed of equal amounts of these four modes, illustrated at four cross sectional planes along the beam. All are in phase at the beam waist.

Although all the modes in a beam share common ω and R values, they *can* be used to describe a beam whose phase front is *not* simply spherical. This is because we are free to alter the phase relationship of the modes—as well as their relative magnitudes—in order to fit the resulting pattern and to match that which exists at a given plane.

It may also be seen that the 'far-field' ($\hat{z} = \infty$) pattern has a uniform spherical phase distribution. The anomalous phase change between $\hat{z} = 0$ and $\hat{z} = \infty$ (or $\hat{z} = -\infty$) for any mode may be given by the expression

$$\delta(m, n) = \tfrac{1}{2}(m + n + 1)\pi. \tag{3.11}$$

For all the modes which contribute to the fields shown in figure 3.2 $m + n + 1$ is *odd*. Equation (3.11), combined with the uniform phase distribution at the beam waist plane, determines that the far field must also have a uniform phase distribution.

This special result is useful as it applies in many cases of practical interest. For example, beams with circular or square symmetry will be composed

33

only of modes for which both m and n are odd. Because of this we may expect as a general result that a beam of circular or square symmetry with a uniform phase distribution in the beam waist plane must also have a uniform phase distribution in the far field.

A uniform phase distribution at the waist plane also leads to a uniform far-field phase distribution when a beam is composed only of modes for which $m + n + 1$ is *even*. However, this situation does not arise very often in practice.

3.2 Apertures and Beam Truncation

One of the most valuable applications for multi-mode analysis is in understanding the effects of apertures upon a beam. This becomes an important subject when we design compact optical systems as we then need to determine how small the optical elements (lenses, mirrors, etc) can be without the effects of beam truncation by the finite apertures becoming too severe.

Most optical systems employ elements which are circular and the beam axis is generally arranged to pass through their centres. We can therefore regard such an element as being placed within a circular aperture through which the beam must pass. Given the circular symmetry which arises in most cases it is convenient to make use of a cylindrical coordinate system.

The derivation of the mode expressions appropriate for Cartesian coordinates is given in detail in chapter 1. In cylindrical coordinates a similar method may be used to obtain suitable mode expressions. The solutions obtained may usefully be divided into two types. Firstly, the modes ψ_p which have circular symmetry are

$$\psi_p = \frac{1}{\omega} L_p\left(\frac{2r^2}{\omega^2}\right) \exp[-j(kz - 2\pi ft)] \, \exp\left\{j\phi_p + r^2\left[\left(\frac{1}{\omega}\right)^2 + \frac{jk}{2R}\right]\right\} \quad (3.12)$$

where r is the radial coordinate, p is the radial mode number, L_p is a Laguerre polynomial of degree p as defined in appendix 1 and

$$\Phi_p = (2p + 1) \arctan\left(\frac{\lambda z}{\pi \omega_0^2}\right). \quad (3.13)$$

Secondly, modes where the field pattern varies with the angular rotation coordinate θ are of the form

$$\psi_{pi} = \frac{1}{\omega} \sqrt{\frac{p!}{(p-i)!}} \left(r\frac{\sqrt{2}}{\omega}\right)^i L_p^i\left(\frac{2r^2}{\omega^2}\right)$$

$$\times \exp\left\{-j(kz - 2\pi ft) + j\Phi_{pi} + r^2\left[\left(\frac{1}{\omega}\right)^2 + \frac{jk}{2R}\right] + ji\theta\right\}. \quad (3.14)$$

L_p^i is a generalized Laguerre polynomial and i the rotational mode number.

In most cases of practical interest the beams and apertures in compact systems may be assumed to have circular symmetry and the ψ_{pi} modes may be neglected.

In order to illustrate the way in which we may employ Gaussian mode techniques to analyse the effects of aperture truncation, we can use the simple example of a beam with circular symmetry being coupled through a hole of radius a.

We can represent the beam of field Ψ directed onto the aperture plane as a linear combination of ψ_p modes:

$$\Psi = \sum_p E_p \psi_p \qquad (3.15)$$

where E_p represents the magnitude and relative phase of the contribution made by the pth mode. The truncated field distribution Ψ' in a plane immediately following the aperture will be of the form

$$\Psi' = \begin{cases} \Psi = \sum_p E_p \psi_p \\ 0 \end{cases} \text{ for } \begin{cases} r \leqslant a \\ r > a. \end{cases} \qquad (3.16)$$

The total beam power P coupled through the aperture may be obtained from the integral

$$\int_0^\infty \Psi' \Psi'^* 2\pi r \, dr = P \qquad (3.17)$$

which, from equation (3.16), is equivalent to

$$\int_0^a \sum_p \sum_q E_p \psi_p E_q^* \psi_q^* 2\pi r \, dr = P. \qquad (3.18)$$

In chapter 1 we introduced the notation $\langle | \rangle$ to represent the integral over the whole plane normal to the beam axis. Here it is convenient to introduce a similar notation which indicates integration over a *finite* portion of a plane. We may define the notation $\{|\}$ such that

$$\{\Psi | \Psi\} = \int_0^a \Psi \Psi^* 2\pi r \, dr \qquad (3.19)$$

and

$$\{\psi_p | \psi_q\} = \int_0^a \psi_p \psi_q^* 2\pi r \, dr. \qquad (3.20)$$

It is also useful to note that, because Ψ' is zero outside the aperture,

$$\langle \Psi' | \Psi' \rangle = \{\Psi' | \Psi'\} = \{\Psi | \Psi\} \qquad (3.21)$$

and we can substitute any of these integrals for each other whenever it is convenient to do so, e.g. we may replace Ψ' with Ψ in most of the integrals over the aperture or where the field is zero outside this region.

We may write that

$$\{\Psi|\Psi\} = \sum_p \sum_q E_p E_q^* \{\psi_p|\psi_q\} \qquad (3.22)$$

and the total power I incident upon the aperture will be

$$I = \langle \Psi|\Psi \rangle. \qquad (3.23)$$

The fraction of power transmitted through the aperture may then be defined as the *power transmission efficiency* T which may be obtained from

$$T = \frac{P}{I} = \frac{\{\Psi|\Psi\}}{\langle \Psi|\Psi \rangle} \qquad (3.24)$$

which allows us to determine how efficiently power is transmitted through the aperture.

Power passing through the aperture may be coupled into a beam of field Φ which represents the pattern of sensitivity of a given detector. In the absence of an aperture, ideal coupling would occur if $\Phi = \Psi$. However, the aperture alters the field *pattern* at the plane just beyond the aperture. The amount of power P_c coupled between the truncated source beam and the detector beam can be obtained from

$$P_c = |\{\Phi|\Psi\}|^2 \qquad (3.25)$$

provided that the detector beam has been normalized correctly. From this we can say that the power-coupling efficiency N between the beams will be

$$N = \frac{|\{\Phi|\Psi\}|^2}{\langle \Phi|\Phi \rangle \langle \Psi|\Psi \rangle} \qquad (3.26)$$

where we have allowed for the possibility that the detector beam may not be pre-normalized correctly.

In many practical cases we are concerned simply with the use of fundamental mode Gaussian beams. For an input beam of this type we may say that $E_p = 0$ for all values of p greater than zero, and we can write that

$$\Psi' = E\psi_0 \qquad \text{for } r \leq a. \qquad (3.27)$$

The fractional power transmission through an aperture of radius a centred on the beam axis will be

$$T_0 = \frac{\{\psi_0|\psi_0\}}{\langle \psi_0|\psi_0 \rangle}. \qquad (3.28)$$

We have defined $\langle\psi_0|\psi_0\rangle$ to be unity and so it follows that

$$T_0 = \int_0^a \left(\frac{1}{\omega}\right)^2 \frac{2}{\pi} \exp\left(\frac{2r^2}{\omega^2}\right) 2\pi r \, dr. \qquad (3.29)$$

By reference to a suitable table of standard integrals we may find the solution of this integral and obtain the result

$$T_0 = 1 - \exp\left(-\frac{2a^2}{\omega^2}\right). \qquad (3.30)$$

Equation (3.29) allows us to specify the amount of power coupled through the aperture. Usually we wish to determine either how efficiently power is coupled into a given detector beam or what the resulting beam pattern is which radiates from the aperture.

The truncated pattern at the aperture may now be regarded as generating a new multi-mode beam, whose field distribution at the aperture plane is Ψ'. We can therefore define an appropriate set of modes to be such that

$$\Psi' = \sum_q A_q \psi'_q \qquad (3.31)$$

where the ψ'_q are a series of mutually orthogonal Gauss–Laguerre modes which have the beam size ω' and phase-front radius R' at the aperture plane.

The coefficient A_q determines the magnitude and relative phase of the contribution that the qth mode makes to the truncated beam. These coefficients may be calculated using the methods outlined earlier. This produces the result

$$A_q = \{\Psi | \psi'_q\} \qquad (3.32)$$

i.e. for the example under consideration where the input beam is simply a fundamental Gaussian we may write that

$$A_q = \int_0^a \frac{1}{\omega\omega'} L_q\left(\frac{2r^2}{\omega'^2}\right) \exp\left[j\delta_q - r^2(\alpha + j\beta)\right] 2\pi r \, dr \qquad (3.33)$$

where

$$\delta_q = (2q + 1) \arctan\left(\frac{\lambda z}{\pi\omega_0'}\right) - \arctan\left(\frac{\lambda z}{\pi\omega_0}\right) \qquad (3.34)$$

$$\alpha = \left(\frac{1}{\omega'}\right)^2 + \left(\frac{1}{\omega}\right)^2 \qquad (3.35)$$

$$\beta = \frac{k}{2}\left(\frac{1}{R'} - \frac{1}{R}\right). \qquad (3.36)$$

We are often only concerned to establish the amount of power coupled into the fundamental mode, $q = 0$, as this enables us to determine how efficiently power passing through the aperture may be coupled to a detector beam of this simple form. By setting $q = 0$ and substituting for L_0 we obtain

$$A_0 = \frac{2 \exp(j\delta_0)}{\pi \omega \omega'} \int_0^a \exp[-r^2(\alpha + j\beta)] \, 2\pi r \, dr. \tag{3.37}$$

Now

$$\langle \psi_0 | \psi_0 \rangle = \langle \psi_0' | \psi_0' \rangle = 1. \tag{3.38}$$

Hence the power-coupling efficiency N_0 through the aperture between the input and output fundamental modes will be

$$N_0 = |A_0|^2 \tag{3.39}$$

which, by referring to a standard table of integrals, leads to the result

$$N_0 = \frac{4[1 - 2\exp(-a^2\alpha)\cos(a^2\beta) + \exp(-2a^2\alpha)]}{(\omega\omega')^2(\alpha + \beta)^2}. \tag{3.40}$$

The power-coupling efficiency N_0 obtained here includes both the losses caused by the failure of some field to pass the aperture and the losses which arise because the field pattern following the aperture is no longer a simple Gaussian and cannot be ideally coupled into a fundamental-mode output beam.

It is useful to note that, in general, N_0 is *not* maximized by setting ω' equal to ω. This may be seen as an effect of the aperture which tends to 'narrow' the beam width by removing the field beyond $r = a$. When the aperture size a is far larger than the input beam size ω, then the effects of truncation are negligible and optimum coupling will arise when $\omega \simeq \omega'$ and $R \simeq R'$. However, when a is comparable with (or less than) ω, then the beam-coupling efficiency may be improved by altering ω' and R'.

Figure 3.3 illustrates how the fractional power transmission T_0 and fundamental-mode coupling efficiency N_0 vary with the ratio a/ω. N_0 is shown both for $\omega' = \omega$ and for $\omega' = 0.8\omega$. It can be seen that, once a/ω is small enough, the power-coupling efficiency can be improved by reducing ω'. For the graphs shown, it is assumed that $\beta = 0$ (i.e. that the beams have identical phase-front curvatures at the aperture plane).

By their very nature, when designing compact optical systems we usually want to make the optical elements as small as possible. Clearly the choice of beam size will have consequences for the system performance; hence there will be a minimum beam waist size acceptable for a given application. Similarly, the choice of a/ω, the ratio of an optical element's size to the beam size will set the degree to which truncation effects may degrade system performance.

The use of single-mode beams is generally advisable as it eases the design process and usually enables the optical system to work well over a wide frequency range. In practice the fundamental mode is used most often. This is partly because it is relatively easy to produce, but it also proves to be the mode which offers the lowest truncation losses. This is because, as the mode number is increased, there is a tendency for the field pattern to have more of the beam power further from the axis.

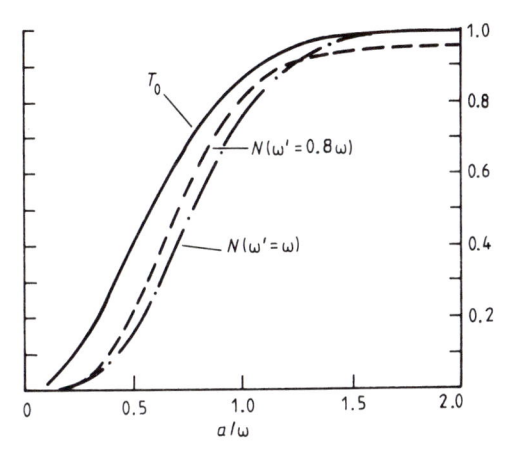

Figure 3.3 Variation in truncation loss T_0 and coupling efficiency N_0 with a/ω for a fundamental Gaussian beam passing through a circular aperture of radius a.

Whilst there is no obvious 'ideal' choice for a/ω, the a/ω value of 1.5 has come to be commonly adopted in practical designs. A single aperture of this size will produce a coupling efficiency N_0 of 0.989, i.e. a beam power loss of -0.05 dB when transmitting a fundamental-mode beam. This is low enough to be insignificant in most circumstances. If a/ω is reduced to 1.3, the coupling loss rises to around -0.15 dB, comparable with the reflection and absorption losses of a typical lens. The losses when $a/\omega = 1.5$ are so low that there is rarely any point in increasing the value significantly. The choice is also convenient in that it gives us a simple 'rule' that lenses, etc, should have a *diameter* of three times the beam size.

3.3 Beam Offsets and Measuring Beam Patterns

A knowledge of how power coupling varies when the beam axes are not coincident is useful when assessing the magnitude of any problems which may arise owing to system misalignments. It can also provide us with a way of measuring the power or field profile of an unknown beam.

For example we can concentrate on the simplest case and evaluate the coupling between two fundamental Gaussian modes whose beam axes are parallel but separated by a distance d.

The beams of fields Ψ and Ψ' can be written as the product of an x- and y-dependent pair of one-dimensional mode expressions, i.e.

$$\Psi = \Psi(x)\Psi(y) \qquad \Psi' = \Psi'(x)\Psi'(y) \qquad (3.41)$$

where $\Psi(x)$, etc, are of the form defined in chapter 1. The total field coupling C between the beams will be

$$C = \langle \Psi | \Psi' \rangle = \langle \Psi(x) | \Psi'(x) \rangle \langle \Psi(y) | \Psi'(y) \rangle. \qquad (3.42)$$

For simplicity we can assume that the axes have been arranged so that the offset d is in the x direction and that the beams are symmetric. We can then say that, ignoring terms which affect only the absolute phase,

$$\langle \Psi(y) | \Psi'(y) \rangle = S \int_{-\infty}^{+\infty} \exp\{-y^2[(\alpha + \alpha') + j(\beta - \beta')]\}\,dy \qquad (3.43)$$

and

$$\langle \Psi(x) | \Psi'(x) \rangle = S \int_{-\infty}^{+\infty} \exp[-x^2(\alpha + j\beta) - (x - d)^2(\alpha' - j\beta')]\,dx$$

$$(3.44)$$

where

$$S = \sqrt{\frac{2}{\pi \omega \omega'}} \qquad (3.45)$$

$$\alpha = \left(\frac{1}{\omega}\right)^2 \qquad (3.46)$$

$$\alpha' = \left(\frac{1}{\omega'}\right)^2 \qquad (3.47)$$

$$\beta = \frac{k}{2R} \qquad (3.48)$$

$$\beta' = \frac{k}{2R'}. \qquad (3.49)$$

Referring to a suitable table of integrals we may find the standard result

$$\int_{-\infty}^{+\infty} \exp[-(ax^2 + 2bx + c)]\,dx = \sqrt{\frac{\pi}{a}} \exp\left(\frac{b^2}{a} - c\right) \qquad (3.50)$$

(provided that $a \neq 0$).

Using this to solve equations (3.43) and (3.44) we obtain

$$\langle \Psi(y) | \Psi'(y) \rangle = \sqrt{\frac{2}{A_y \omega \omega'}} \qquad (3.51)$$

where

$$A_y = (\alpha + \alpha') + j(\beta - \beta') \tag{3.52}$$

and

$$\langle \Psi(x) | \Psi'(x) \rangle = \exp(-d^2 B_x) \sqrt{\frac{2}{A_x \omega \omega'}} \tag{3.53}$$

where

$$A_x = (\alpha + \alpha') + j(\beta - \beta') \tag{3.54}$$

and

$$B_x = (\alpha' - j\beta')\left(1 - \frac{\alpha' - j\beta'}{A_x}\right). \tag{3.55}$$

By substituting equations (3.51)–(3.55) into equation (3.42) we can now define how the field coupling C between the two beams will vary as a function of the offset distance d.

In practice we often wish to check that the beam waist location and size of a beam are correct. This can be done by employing a detector with an antenna which responds to a known beam pattern and observing how the coupled field magnitude and phase vary as the detector position is altered. The expressions given above then enable us to use the measurements to establish beam waist parameters. Unfortunately, we are frequently only able to employ a power detector which cannot provide us with information on the relative phase variations as we move around the beam that we wish to measure.

The power-coupling efficiency $N(d)$ between the offset beams may be calculated from

$$N(d) = |C|^2. \tag{3.56}$$

In principle we can then demonstrate that, if we measure how $N(d)$ varies at three separate planes, we can deduce both the beam waist size and the beam location. Hence we can choose to make our measurements at any three suitable planes.

If we can predict the beam size and location, then we expect this method can be used fairly easily to confirm or contradict our prediction. This can be done because firstly the size of the beam varies symmetrically on both sides of the beam waist plane and secondly, at the beam waist plane, $\beta = 0$.

If we place our power detector, whose beam waist size is ω'_0, such that its waist plane is coincident with that of the beam which we wish to measure, then at the common waist plane we can calculate that $N(d)$ will be

$$N(d) = \frac{4 \exp(-2d^2/\varsigma^2)}{(\omega'_0/\omega_0 + \omega_0/\omega'_0)^2} \tag{3.57}$$

where ω_0 is the beam waist size of the detector beam or antenna, and

$$\zeta^2 = \omega_0^2 + {\omega_0'}^2. \tag{3.58}$$

From this result we can see that the measured pattern of $N(d)$ has a Gaussian profile. Knowing ω_0' we can use the measured value of ζ to obtain the beam waist size ω_0.

To check that we *have* correctly identified the beam waist plane, we need to measure only the beam widths at two other planes, spaced equal distances along the beam on either side of the waist. From the symmetric nature of a Gaussian beam these other planes should produce beam sizes which are firstly larger than ζ and secondly equal to each other. If this is not the case, then we did *not* initially have the waist planes collocated. If we can rely upon the absolute calibration of the power detector and the total beam power, the relative sizes of ω_0 and ω_0' can also be deduced from $N(0)$.

Chapter 4

Antennas and Feed Systems

In the first three chapters of this book we have established how we can deal with transmitting signals from place to place using beams in free space. Many of the actual signal sources, detectors, etc, are mounted in a length of metal waveguide. We therefore require some way of coupling signal power back and forth between free-space beams and waveguides. We need some form of *antenna* or *feed system* to transform one sort of mode into another.

Figure 4.1 illustrates one of the most common arrangements used for millimetre-wave devices. The arrangement represents how power in a free-space beam may be coupled into a small device such as a detector diode mounted in a short length of metal waveguide. The manner in which signal power is coupled from a waveguide into a device will be discussed in a later chapter. Here we shall concentrate on how power is transferred from free space into (or out of) a waveguide.

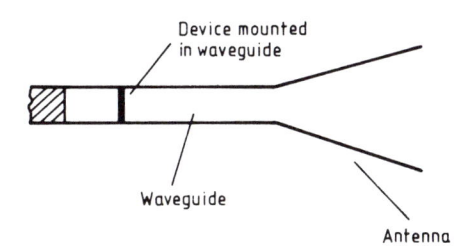

Figure 4.1 Combination of antenna, transmission line and back-short arranged to couple a device to an electromagnetic signal propagating in free space.

4.1 Basic Coherent Antenna Theory

Before analysing the properties of millimetre-wave systems in terms of Gaussian modes it is useful to establish some of the general properties of antennas. An antenna may be regarded as a sort of 'transformer', taking one form of field pattern and converting it into another. When a number of the basic properties are looked at in this way, they become fairly clear.

Firstly, an antenna cannot of itself produce any 'extra' signal power. Even a perfect antenna can only output the same amount of signal power that we initially put in. For any real antenna, of course, some signal power will be 'lost'; for example, currents set up in metal parts will cause signal power to be wasted in warming up surface resistances.

Secondly, an antenna is a *reciprocal* system, i.e. it works in much the same way irrespective of which way we pass signal power through it. To see just what this means, consider the arrangement shown in figure 4.1 but with the detector replaced by a signal source. The source will produce a Ψ_g mode pattern which propagates along the guide to the antenna. The antenna then transforms Ψ_g into a Ψ_s pattern which it radiates out into space.

Imagine taking a video of the field pattern moving through the system. If we replay the video *backwards*, the field Ψ_s appears to enter the antenna from free space and is transformed into the Ψ_g pattern moving along the guide away from the antenna. For an antenna dealing with a coherent signal we cannot tell by looking at the video which way we are running the playback! This is a very important feature of antennas. It means that the only real difference between a 'transmitting' antenna (one used to radiate a signal out into space) and a 'receiving' antenna (used for collecting power from free space) is the direction that we have chosen to pass signal power through the system.

The reciprocal nature of antennas means that we often obtain good results by using essentially identical antennas to radiate and receive beams. It also turns out to be very helpful when we analyse or design antennas. This is because we can often choose to carry out our analysis by thinking of the antenna as either transmitting *or* receiving a signal irrespective of its actual role.

Figure 4.2 illustrates one of the simplest possible forms of antenna, the *Hertzian dipole*. Antennas of this form are not normally employed for millimetre-wave systems, but it is a useful example which serves to establish some basic antenna properties.

We can inject a signal at some chosen frequency into the antenna via the pair of parallel wires connected to the break near its centre. The actual antenna consists of two short (compared with the signal wavelength) pieces of wire and the signal sets up a uniform alternating current on the antenna. In order for the charge moving at the wire ends to have 'somewhere to go', we need to arrange for some capacitance between the wire ends. This can be

Antennas and Feed Systems

done by attaching the ends to some large discs, spheres, etc. The electric field between their end areas then acts as a capacitor into which end current may flow.

For simplicity we shall ignore any effects of the end shapes upon the antenna and regard the arrangement as simply a length L of wire upon which we have set up a uniform alternating current

$$I(t) = I_0 \sin(\omega t). \tag{4.1}$$

By reference to a suitable book on electromagnetism it will be found that this current will generate an electric field E given by the expression

$$E = \frac{60\pi LI \sin\theta \cos(\omega t - kr)}{\lambda r} \tag{4.2}$$

where r is the distance from the dipole centre to the point where we wish to determine the field E, and θ is the angle between the dipole wire and the line connecting the point and the dipole centre.

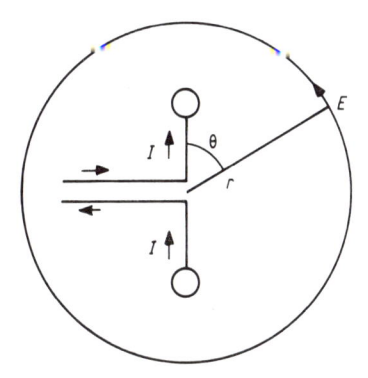

Figure 4.2 Electric field E at a distance r and direction θ radiated by a Hertzian dipole carrying a current I.

The root-mean-square field at the point (r, θ) will therefore be

$$E(r, \theta) = \frac{60\pi LI \sin\theta}{\sqrt{2}\lambda r}. \tag{4.3}$$

If we imagine the dipole as being placed at the centre of a sphere of radius r, then the power $P(\theta)$ per unit area passing through the surface of the sphere around the point (r, θ) will be

$$P(\theta) = \frac{E^2(\theta)}{Z} \tag{4.4}$$

where Z is the impedance of free space, i.e. we may say that

$$P(\theta) = \frac{(60\pi)^2 L^2 I^2 \sin^2 \theta}{2Z\lambda^2 r^2}. \qquad (4.5)$$

The way in which P varies with direction is generally referred to as the *power pattern* or *antenna pattern* of the antenna. In a spherical coordinate (r, θ, ϕ) system the power pattern often depends upon *both* angles and may be given as an appropriate function $P(\theta, \phi)$. In this particular case the pattern has rotational symmetry in the ϕ direction, i.e. the power does not vary when we move around the antenna provided that we keep θ constant.

Some workers also refer to $E(\theta, \phi)$ as the antenna pattern. To avoid confusion here we shall use the term *power pattern* for $P(\theta, \phi)$ and *antenna pattern* for $E(\theta, \phi)$.

From equation (4.5) we can see that the power pattern is *directional*, more power being transmitted in some directions than others. In particular, *no* power is transmitted along the two directions $\theta = 0$ and $\theta = \pi$. Antenna engineers often invoke an *omnidirectional* antenna—the *isotropic radiator*—which is assumed to radiate power equally in all directions. The behaviour of a particular arrangement may then be compared with an isotropic antenna.

Consider a pair of antennas, one a Hertzian dipole and the other omnidirectional, which are used in turn to radiate the same total power out into space. The Hertzian dipole will radiate *less* power in some directions (those around $\theta = 0$ or π) than the omnidirectional antenna. Since the total radiated powers are equal, it follows that *more* power is transmitted by the dipole in some other directions (in fact, in the directions around $\theta = \frac{1}{2}\pi$).

If we were to set up a detector some distance from an omnidirectional antenna and measure what happens when it is replaced by a Hertzian dipole, we find that—on the $\theta = \frac{1}{2}\pi$ plane—the received signal increases by around 50%. So far as the detector is concerned, the result is no different from if we had amplified the transmitted power. From this has grown the concept of antenna *gain*.

We can define the *gain* of an antenna as being the factor by which we have increased the received signal by changing the antenna from some standard type (e.g. isotropic). When using the term 'gain', however, it is important to note that the total power radiated has *not* been increased. Instead, we have arranged not to 'waste' power by transmitting in directions where it will not be received. Because of this the gain and the *directionality* of an antenna are closely linked.

In fact, if we are dealing with a single coherent mode, it can be shown that it is impossible to produce a truly isotropic radiator. This may be seen by appeal to a rule from topology: 'You can't comb a hairy billiard ball smooth!' The coherent wave radiated from an antenna will have its electric field perpendicular to its direction of propagation. If we consider the field moving out through a large sphere centred on the antenna, we find that the power is moving radially. Hence the field electric vector is everywhere parallel to the local surface of the sphere.

In order to satisfy Maxwell's equations in free space, we cannot let the field distribution have any discontinuities, i.e. the field vector direction and magnitude must vary smoothly from place to place on the sphere. We can then think of the field as like the hairs on a hairy billiard ball. No matter how they are combed, there will always be a tuft somewhere and a gap somewhere else. This would represent a discontinuity as the field direction (e.g. the direction in which the hairs point) changes abruptly as we move through the tuft or gap. To avoid this, we can arrange for the magnitude of the field (e.g. the length of the hairs) to fall smoothly to zero as we approach either the tuft or the gap.

When dealing with a vector field where magnitude, direction (or polarization) and phase are uniquely defined, we cannot therefore make a real isotropic radiator. However, we *can*—in principle at least—make one if we are only concerned with power.

For example a single dipole will transmit no field in the directions along its axis. We can, however, use a pair of crossed dipoles and supply them with signal at the same frequency, but $90°$ out of phase. Now we are essentially radiating *two* fields which are orthogonal in phase. The 'gaps' in each pattern are filled by radiation from the other dipole. The power pattern is now much more even than for a single dipole. The radiated field of such a pair is *elliptically* polarized. The precise polarization state of the wave that we observe depends upon the angle between the local wave propagation direction and the plane of the two dipoles.

Despite these difficulties, the gain of an antenna is often quoted as being so many decibels compared with an (assumed) isotropic radiator. This gain is, in practice, normally determined by measuring the *directional* pattern of the antenna or by comparing it with a 'standard' antenna whose gain or pattern are taken as 'known'.

Figure 4.3(*a*) shows a possible power pattern for a 'high-gain' antenna. A real pattern will be fairly complex. Most of the transmitted power will be beamed into a fairly narrow solid angle which is conventionally called the *main lobe*. The remaining power being sent in other directions into what are called the *side lobes*. The detailed analysis of such a complex pattern is quite difficult. Fortunately, for most purposes we can replace it with a much simpler 'idealized' antenna pattern. Figure 4.3(*b*) illustrates an 'idealized' pattern. Note that both patterns only show how the radiated power varies with *one* of the coordinate angles. Strictly speaking we should picture the variations in power with both ϕ and θ. For many antennas the actual pattern can be assumed to have circular symmetry about the axis of the main lobe; hence only one 'slice' through the pattern need be shown.

The main lobe of the real pattern is taken to be the solid angle within the power minima which surround the direction of maximum gain. We can calculate the amount of power radiated in the main lobe by integrating $P(\theta, \phi)$ over this solid angle, i.e. we can say that

$$\int_{\text{Main}} P(\theta, \phi) \, d\Omega = P_M \Omega_M \tag{4.6}$$

where P_M is the maximum value of $P(\theta, \phi)$ and the integral is over the main lobe only. By defining the *normalized* power pattern

$$P_n(\theta, \phi) = \frac{P(\theta, \phi)}{P_M} \tag{4.7}$$

this can also be written in the form

$$\int_{\text{Main}} P_n(\theta, \phi) \, d\Omega = \Omega_M. \tag{4.8}$$

This lets us define the solid angle Ω_M of an idealized main lobe which would radiate the same total power but have a *uniform* value of $P(\theta, \phi)$ equal to the *maximum* value P_M of the real antenna. The value Ω_M is conventionally referred to as the *main beam solid angle*. Any remaining power, radiated in the side lobes of the real antenna, is assumed for the idealized pattern to be radiated uniformly in all directions *outside* Ω_M.

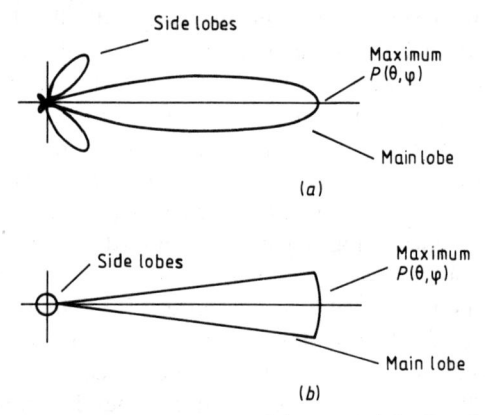

Figure 4.3 Comparison of a 'real' (a) and 'idealized' (b) power pattern of a directional antenna.

It is also useful to define another conventional quantity Ω_A, the *antenna solid angle*, via the expression

$$\int_{4\pi} P_n(\theta, \phi) \, d\Omega = \Omega_A. \tag{4.9}$$

Ω_A is the solid angle into which *all* the output power would be directed if the pattern were uniform and *no* power was radiated into the side lobes. Hence we can see that, for an antenna without side lobes, $\Omega_A = \Omega_M$.

Power radiated into the side lobes is generally wasted. Hence most antennas are designed to minimize their side lobes and, in practice, we can often assume that Ω_A and Ω_M are effectively the same.

The antenna gain only arises because power is directed into a limited solid angle Ω_A instead of being radiated into 4π sr. From the solid geometry of the radiation pattern we can observe that the gain G of the idealized antenna pattern compared with an isotropic radiator will be such that

$$G = \frac{4\pi}{\Omega_A}. \tag{4.10}$$

Hence we need only be given either the antenna solid angle or the gain to be able to calculate the other value.

Consider now a receiving antenna, placed so that it intercepts a field E over a collecting area A. The received power will be

$$W = \frac{|E|^2 A}{Z} \tag{4.11}$$

where Z is the impedance of free space.

Earlier in this chapter it was pointed out that the distinction between a receiving and transmitting antenna is solely one of use, i.e. in which direction the field passes through the antenna. We can therefore expect that there should be some connection with the 'collecting area' of an antenna when used to receive radiation and its 'gain' or 'directionality' when used to transmit. By reference to a suitable book on antennas it will be found that the effective area A of an antenna and its antenna angle Ω_A are related via the expression

$$\lambda^2 = A\Omega_A \qquad \Omega_A = \frac{\lambda^2}{A} \tag{4.12}$$

where λ is the radiation wavelength. Once the wavelength is known, the main properties of the antenna are defined by just *one* of the three parameters, A, Ω_A or G. The other two may be then be calculated if required.

When a system of the type illustrated in figure 4.1 is used with a waveguide-mounted source a field Ψ_g in the waveguide will be transformed into a field Ψ_s radiated into space. If only one mode pattern for Ψ_g may propogate along the guide at this frequency, the pattern radiated into space is uniquely defined. The antenna is a reciprocal system. When used as a receiver it will be sensitive to the specific field pattern Ψ_s in free space. Any power that arrives which is in this particular pattern will be efficiently coupled into the waveguide.

Any field pattern Ψ in the region surrounding the antenna may be thought of as being a sum of two orthogonal components Ψ_s and Ψ_o. Power in the field Ψ_s is transformed into the guide field Ψ_g pattern. Power in the orthogonal field Ψ_o will attempt to produce some other pattern in the guide. However, we have arranged that only *one* guide mode is possible. Hence any power in the field Ψ_o will be rejected by the antenna.

Communications engineers, radio astronomers, etc, use antennas to detect signals which are in the form of plane waves radiated by a distant source. Under these circumstances the main concern is to arrange for an antenna pattern which responds as strongly as possible to a plane-wave field coming from the direction of the source of interest, i.e. the antenna field Ψ_s should be as close as possible to a plane wave.

In compact optical systems where we are dealing with beams, we would wish Ψ_s to be as similar as possible to the signal field of the beam that we wish to receive. Radio astronomers, etc, may also wish to receive power from an 'extended' source, i.e. one which has a measurable angular extent in the sky. Here the ideal antenna field may be designed to respond over a specific limited solid angle.

In each case the efficiency of power coupling through the antenna will depend upon what fraction of the incoming field Ψ is in the pattern Ψ_s to which the antenna will respond. When dealing with plane waves or with fields from a source of limited angular size, it is very convenient to treat the antenna in terms of its effective area, or gain, or directionality. When dealing with beams, however, we often find that we need to employ the arguments of field coupling outlined in the first three chapters.

The effective antenna area that we have defined may not always be clearly identified with a specific physical area. Consider, for example, a dipole. The actual surface area of the dipole wires is very small and varying the wire diameters to alter their area has little or no effect upon the antenna's behaviour. For a dipole there is no clear physical area which can be seen to relate to the antenna's 'effective' area.

This having been said, for many of the most common forms of millimetre-wave or microwave antennas the physical areas are reasonably large compared with λ^2 and the antenna effective area tends to be at least similar to an identifiable physical area.

4.2 Feed Horns

In contrast with the dipole, most millimetre-wave antennas can best be analysed as an aperture which frames a defined field pattern. This means that we can usually identify a physical area which is closely related to the effective area A.

Perhaps the simplest form of antenna that we can use with a metal waveguide is to leave the end of the guide open to free space. The signal travelling along the guide produces a field pattern at the open end which we

Antennas and Feed Systems

can define by knowing the details of the signal modes in metal waveguides. We can then treat the open end as being a rectangular aperture.

For a standard guide carrying just the fundamental TE mode, the electric field will be

$$E = \begin{cases} E_0 \sin\left(\dfrac{\pi x}{a}\right) \exp(j\omega t) & \text{within the aperture} \\ 0 & \text{outside the aperture} \end{cases} \quad (4.13)$$

the electric field being plane polarized with its vector parallel to the short wall of the guide. The x direction is assumed to be parallel to the longer wall which is of length a.

For a standard guide the width and height will be around $\frac{1}{2}\lambda$ and $\frac{1}{4}\lambda$; hence the effective area will be approximately $\frac{1}{8}\lambda^2$. Using equation (4.12) we can see that this implies an antenna angle Ω_A of about 8 sr, i.e. about three quarters of a complete sphere! Clearly, in order to produce a reasonably directional beam, the effective area of the antenna needs to be somewhat larger than $\frac{1}{8}\lambda^2$. Hence an open waveguide end is unsatisfactory for most purposes.

The simplest way to obtain a large area is to flare out the part of the waveguide near the open end into a pyramidal shape. A wave travelling along the guide sees this as a length of guide whose cross sectional area increases gradually as it moves towards the open end. This has the effect of 'stretching out' the field pattern without altering its shape. If required, the angles may be different for the two pairs of walls and we can, for example, have a square opening at the 'base' of the pyramid.

In this way we can produce an open end which is, say, $5\lambda \times 5\lambda$ across. If we input the fundamental TE mode the field at the open end will be

$$E = E_0' \sin\left(\dfrac{\pi x}{a'}\right) \exp(j\omega t) \quad (4.14)$$

at an aperture of width a'. Note that the overall field magnitude must fall as the total power leaving the open end cannot exceed what we put in via the waveguide. For an aperture of area A and a waveguide of area $\frac{1}{8}\lambda^2$ conservation of power allows us to write that

$$E_0' = \dfrac{E_0 \lambda}{\sqrt{8A}}. \quad (4.15)$$

Taking the example of $A = 25\lambda^2$ we obtain an antenna angle $\Omega_A = \frac{1}{25}$ sr, i.e. if the main lobe was symmetric it would be about 12° across in any direction. This is directional enough to be coupled effectively into a beam.

Antennas of this general type are called *feed horns* partly as a reference to their shape and partly because they are used to 'feed' radiation from one place to another.

ne care must be taken when designing or building feed horns which have a very steep flare angle or which have a large open end. As a field moves closer to the open end, the effective waveguide size increases and the cut-off frequencies for higher TE (or TM) modes will fall below the signal frequency. Any small imperfections in the horn may then alter the field pattern, changing the behaviour of the antenna. Also, the field moving within the horn does *not* have a plane phase front (otherwise it would not spread out!). The mode inside the horn will tend to have a spherical phase front centred on the horn apex. Hence there may not be a perfect transfer of signal between the guide and horn modes. This problem becomes more severe if the horn angle is increased.

In practice, the description of horn behaviour given above is fairly reliable provided that

(i) the aperture is no more than a few wavelengths across and
(ii) the flare angle is not more than around 10–20°.

Beyond these limits the analysis and design of feed horns should be undertaken using a more involved method. Fortunately, we can normally make feed horns offering acceptable beam-coupling performance within these limitations.

4.3 The Scalar Feed

Having obtained a suitable antenna area, gain and directivity, we ideally require an antenna which produces (or is sensitive to) a single free-space mode—preferably the fundamental Gaussian mode with circular symmetry. For the fundamental TE mode, the field at the aperture of a pyramidal horn varies sinusoidally with x but does not alter as we move in the y direction (unless we move outside the aperture!). Hence the pattern radiated by such a feed cannot have the same form in the x and y planes—even if the aperture is square.

If we require a free-space pattern which is simply a fundamental Gaussian, we need to generate a Gaussian field pattern at the horn aperture. Various methods have been employed to achieve this. One of the most successful arrangements is called the *scalar* or *conical corrugated feed*. This differs from the pyramidal horn in two ways.

(i) As the guide size is gradually increased, it is changed from rectangular to circular cross section.
(ii) A series of radial corrugations are introduced into the metal walls.

The detailed analysis of the scalar feed is quite complex. For our purposes, however, it is sufficient just to give an outline of the general behaviour.

For a TE mode in a rectangular guide we can expect that, just above the surface of a metal wall, the E-field component parallel to the surface and perpendicular to the axis must be zero. However, the E-field component perpendicular to the surface can be just as large as it is further towards the centre of the guide. At points a half-wavelength apart *along* the guide the perpendicular E fields will be out of phase. Hence there will be a potential difference along the guide wall which sets up a current flow along the guide, parallel to the axis of propagation.

Consider what would happen when we cut a series of narrow grooves, each a quarter-wavelength deep, into the guide walls. The grooves are placed so that current wanting to flow along the guide is forced to travel down the groove, across the bottom and back up the other side. Because of the time that this takes, the currents at the two edges of a groove are $180°$ out of phase. Although the currents are forced to go 'the long way round', the groove edges are very close to each other. As a result, any fields set up at a groove edge are effectively cancelled by an equal and opposite adjacent field. It becomes impossible to set up a potential difference or current along the wall.

The effect of the grooves is to 'kill' the E field near the wall surface which cannot now have a non-zero component perpendicular to the surface. The field pattern in the guide must now fade gracefully to zero as we approach *any* of the walls.

If we wish, we can make a square pyramidal horn with corrugated walls. This can be used to produce an aperture pattern which varies symmetrically. The resulting pattern does not, however, radiate a simple fundamental Gaussian. Instead it produces a multi-mode beam with much of the power spread over the first few modes. By altering the waveguide profile to circular, we can change the field profile into a shape which is a close approximation to a Gaussian.

Figure 4.4 illustrates a typical conical corrugated feed. For a real feed horn of this type it is useful to introduce two extra features not mentioned above. Firstly, the conversion from rectangular to circular guide takes place *before* the guide is flared out into a cone. Secondly, the first few grooves at the 'throat' of the horn (i.e. the end nearest the waveguide) are different in length from the rest. Both of these features make the transition from the normal rectangular TE mode to the field pattern in the corrugated horn more gradual. If the change from rectangular guide to corrugated cone is too abrupt, the field patterns do not couple efficiently into one another and a significant amount of signal power may be reflected.

A detailed analysis of this type of horn reveals that the field at the aperture can be well described as a spherical 'cap'. The radius of curvature R of the phase front which passes through the aperture rim is at the cone apex. It may be shown that the field distribution at the aperture is plane polarized and of the form

$$E_y = A J_0^T(\alpha r) \exp\left(-\frac{jkr^2}{2R}\right) \exp(j2\pi ft) \quad (4.16)$$

where

$$\alpha = \frac{2.405}{a} \quad (4.17)$$

$$k = \frac{2\pi}{\lambda} \quad (4.18)$$

and

$$J_0^T(\alpha r) = \begin{cases} J_0(\alpha r) \\ 0 \end{cases} \text{if} \begin{cases} r < a \\ r \geqslant a \end{cases} \quad (4.19)$$

where J_0 is a Bessel function of the first kind and zeroth order and a the aperture radius. $J_0(2.405) = 0$; hence the distribution is arranged so that the field falls smoothly to zero at the aperture rim.

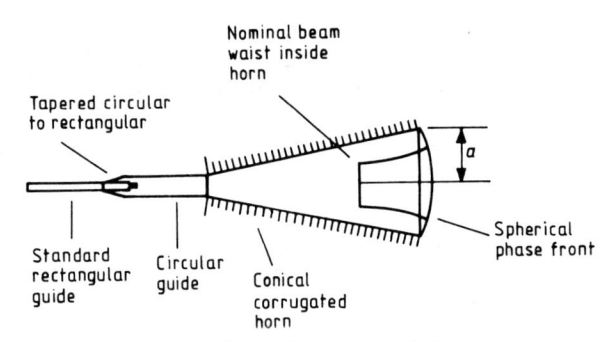

Figure 4.4 Conical corrugated feed horn.

By using the methods outlined in chapter 3 it is possible to define a series of Gauss–Laguerre free-space modes which would produce the aperture distribution of equation (4.19). From this analysis we find that the aperture field may be represented by

$$J_0^T(\alpha r) = \sum_p A_p \frac{1}{\omega} L_p\left(\frac{2r^2}{\omega^2}\right) \exp\left(-\frac{r^2}{\omega^2}\right) \quad (4.20)$$

where ω is the beam size which at the aperture plane may be taken from

$$\omega = 0.6435 a. \quad (4.21)$$

The power radiated in the pth mode will be $|A_p|^2$. If we arrange that the

Antennas and Feed Systems

total radiated power P_t is unity, i.e. if

$$P_t = \sum_p |A_p|^2 = 1 \tag{4.22}$$

then it can be shown that

$|A_0|^2 = 0.9792$ $|A_1|^2 = 4.9 \times 10^{-9}$
$|A_2|^2 = 1.45 \times 10^{-2}$ $|A_3|^2 = 1.86 \times 10^{-3}$
etc.

From this result it can be seen that around 98% of the power radiated by the conical corrugated feed will be in the fundamental Gaussian mode. For many purposes we may regard such a feed horn as an ideal antenna for coupling between a waveguide device and a free-space Gaussian beam. From the beam size ω and the phase-front radius of curvature R at the aperture we can also work out an effective beam waist radius and location for the antenna beam produced by the horn. It is then normally convenient to regard the horn performance as depending upon this waist size and position.

The main problems of conical corrugated horns are consequences of the grooved structure which produces their highly useful antenna pattern. The grooves must be around one quarter of a wavelength deep. They must also be fairly narrow and closely packed. Millimetre-wave horns are usually produced by electroforming. A steel or aluminium 'mandril' is turned on a lathe. The actual horn is then formed by copper electroforming onto the outside of the mandril.

When making a 'smooth' horn the mandril may be pulled out and re-used. This is impossible when the mandril is grooved as this locks it into the grown copper. Some manufacturers have used a technique of combining a smooth metal horn with a plastic (usually injection moulded) pattern for the grooved outer part. The horn is pulled out and the plastic then dissolved by a suitable solvent. An alternative approach is to use an aluminium mandril which is then eaten out using a reactive liquid which attacks aluminium but not copper. This produces excellent results but means that a new mandril must be made for every horn.

Above around 300 GHz (1 mm wavelength) it becomes difficult to machine and form the required grooves as they are becoming too small to cope with accurately. Hence corrugated feeds are rare above 200 GHz.

The use of quarter-wavelength grooves also 'tunes' the feed horn, i.e. we can only obtain the desired behaviour at signals whose wavelength is appropriate for a given horn. In practice this tends to mean that corrugated feeds work well over about a 20% frequency range. Outside this range the antenna pattern deteriorates. The grooves also tend to make the horn couple less well to the waveguide at frequencies well away from the 'design' frequency. Hence the signal losses also increase outside a narrow range.

Perhaps ironically, the imperfections of a smooth metal horn can sometimes mean that they work *better* than a corrugated one at high frequencies! At frequencies around 300 GHz and above the losses due to surface resistance in a copper horn become significant. This tends to absorb signal power from the field near the walls of the horn. The result may be to produce an aperture field pattern which is closer to Gaussian than would otherwise occur. Hence the antenna pattern of a simple smooth horn may be quite useful although the result is obtained by a process of selective signal loss.

4.4 'Vivaldi' and Other *E*-plane Antennas

A number of alternatives to the conical corrugated feed have been proposed, each offering a particular combination of good and bad points. Rather than spending time listing them and discussing their relative merits it is more profitable to compare the general idea of feed horn 'aperture' antennas with another class of antennas based upon a different sort of waveguide.

The behaviour of conical and rectangular feed horns can be described by considering them as tapered lengths of a standard metallic waveguide. In a similar way the properties of another class of antenna can be obtained from their similarity to the *E-plane guide*.

Figure 4.5 illustrates an *E*-plane waveguide linked to two slightly different antenna arrangements. The guiding structure consists of a narrow gap between the parallel edges of two metal plates. A signal may be transmitted along the gap, power being carried by the electromagnetic field which surrounds it. The properties of this type of guide are almost the same as those of the *twin feeder* which will be discussed in the next chapter.

An *E*-plane guide is often mounted on a dielectric substrate in order to give it stability and strength. This will, however, affect the field pattern near the gap. A detailed analysis of the arrangement indicates, that, provided that the thickness t of any dielectric substrate satisfies the condition

$$t \leqslant \frac{\lambda}{4(\varepsilon - 1)} \qquad (4.23)$$

and the gap width is less than a half wavelength, then power may be transmitted in a single TEM mode.

When discussing rectangular waveguides it was pointed out that their modes are of two types: TE modes where the electric field is wholly transverse; TM modes where the magnetic field is wholly transverse. For an electromagnetic wave in free space *both* the electric field and the magnetic field are wholly transverse. This property is shared by the E-plane mode of interest here. TEM stands for *transverse electric and magnetic* fields.

Just as was the case with a rectangular waveguide we may produce a form of antenna by flaring open the gap before terminating the guide at an open end. Provided that the flare angle is gentle, the field pattern 'stretches' smoothly until it covers a significant area by the time that it arrives at the end of the guide. Although the system differs from rectangular guides in that there is no obvious physical aperture, we can still define the field pattern at the final plane and use this to determine the pattern radiated out into free space.

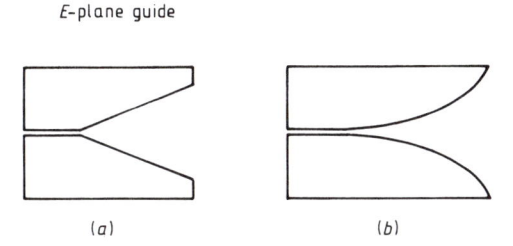

Figure 4.5 Linear (a) and 'Vivaldi' (exponential) (b) planar antennas.

Two examples of E-plane antennas are shown in figure 4.5. The *linear taper slot antenna* is shown in figure 4.5(a). This is a direct analogue of a rectangular waveguide horn, tapered or flared at a fixed angle. The *Vivaldi antenna* shown in figure 4.5(b) differs in that the flare angle varies gradually as we move along the antenna. In a linear taper slot antenna arrangement the gap increases linearly as we move away from the guide. In the Vivaldi antenna the gap increases exponentially.

E-plane guides and antennas are planar and can be manufactured by photolithography and printing processes very similar to those used to make ordinary electronic printed-circuit boards. This opens up the possibility of making large numbers of identical systems relatively cheaply. It is also possible to produce much finer detail by photolithography than by normal machining. Hence these systems may prove to be more useful at very high frequencies where rectangular waveguide and corrugated feeds are almost impossible to make.

Another advantage of E-plane antennas is related to the nature of TEM waves. It is a general property of the fundamental TEM mode that it does not cut off at any signal frequency above zero. We can exploit this to help to produce antenna systems which work well over a very wide frequency range.

Consider what would happen if we made an antenna which had a very large open end and tapered smoothly down into an E-plane guide which had a very narrow gap. Over a fairly large frequency range the signal wavelength would remain large compared with the gap and small compared with the opening. Hence the behaviour of the antenna will depend more upon the taper angle or shape than upon the sizes of the ends. As the input guide gap can be made quite small without having to worry about low frequencies being cut off, it becomes possible to make antennas which work well over wide frequency ranges.

4.5 Behaviour of a Cassegrain System

From the field Ψ produced in a plane near an antenna we can obtain appropriate values of the beam waist size, the location and the mode composition of the beam. Using the methods outlined in this and previous chapters we can then calculate the field pattern at any plane from the antenna out to the far field (i.e. where z approaches infinity).

As we have already discovered, the behaviour of an antenna is to a large extent determined by a simple measurement of the antenna or beam solid angle. Given the beam waist size ω_0, we can immediately calculate the angular extent of the antenna beam in the far field.

From equation (1.13) the beam size ω at a distance z from the beam waist will be such that

$$\omega^2 = \omega_0^2 \left[1 + \left(\frac{\lambda z}{\pi \omega_0^2}\right)^2\right]. \tag{4.24}$$

Hence, when z is very large,

$$\omega \simeq \frac{\lambda z}{\pi \omega_0} \tag{4.25}$$

and we can define a beam divergence angle θ_0 in terms of the differential:

$$\theta_0 = \frac{\delta \omega}{\delta z} \simeq \frac{\lambda}{\pi \omega_0}. \tag{4.26}$$

This angle may then be taken as a measure of the effective beam size in the far field. The beam *solid* angle of a beam of circular symmetry, given by $\Omega_0 = \pi \theta_0^2$, will then be equal to

$$\Omega_0 = \frac{\lambda^2}{\pi \omega_0^2} \tag{4.27}$$

which has a similar form to equation (4.12) linking an antenna's beam angle with its effective aperture area. A circular disc of radius ω_0 would in fact have an area $\pi \omega_0^2$; hence it is convenient to identify ω_0 as the radius of the antenna's effective area.

For a pure fundamental mode beam the field amplitude falls to $1/e$ of its axial value when we are offset by ω from the beam axis, i.e. the beam solid angle Ω_0 represents the beam size to the $1/e^2$ power positions. Choosing a different measure for the beam size (e.g. the half-power size) would have produced a different measure of the far-field beam solid angle.

When we are discussing fields which are intended to produce a Gaussian beam, the beam size is often given as an angle measured to the $1/e$ *power* positions in the far field. Alternatively, fields are sometimes described in terms of an *F number*. This usage differs from the more common one of the '*F* number' of a lens or mirror.

The $1/e$ power width of a fundamental Gaussian is $\omega/\sqrt{2}$ and the far-field power half-angle θ_p will be

$$\theta_p = \frac{\lambda}{\sqrt{2}\pi\omega_0}.$$

For such a horn, the *F number* is often defined as the ratio of the total width $2\omega/\sqrt{2}$ of the beam between the $1/e$ power points in the far field to the distance from the beam waist, i.e.

$$F = \frac{z}{\sqrt{2}\omega} \qquad (4.29)$$

or

$$F = \frac{2}{\theta_p}. \qquad (4.30)$$

The properties of a system may be defined in terms of any of the above quantities. Provided that we keep clearly in mind which method is being used, any of the measures mentioned above may be employed.

Figure 4.6 represents a Cassegrain antenna system. The beam produced by such a system can be described in terms of the field pattern produced in an appropriate aperture plane.

For simplicity we may choose either the primary or the secondary mirror locations as the nominal aperture plane. The secondary mirror can then be treated as a 'centre stop' which blocks the middle portion of the aperture field. The primary size defines an aperture which truncates the field pattern. The aperture field Ψ can then be described in terms of an appropriate Gaussian mode set.

In reality the primary and secondary mirrors will be curved and hence have finite depths; they will also be in distinct planes. However, in general, the effective beam waist size ω_0 will be sufficiently large that $\lambda Z/\pi\omega_0^2 \ll 1$ for any of these distances. Hence the assumption of a single aperture plane is good for most purposes.

For example we can consider the case where the antenna is coupled to a feed which produces a fundamental-mode Gaussian field pattern at the aperture plane which has a uniform phase distribution, i.e. $R = \infty$, and the aperture is at the waist plane of the projected beam.

For a circularly symmetric antenna system the resulting aperture field will be of the form

$$\Psi = \begin{cases} \dfrac{1}{\omega_0} L_0\left(\dfrac{2r^2}{\omega_0^2}\right) \exp\left(-\dfrac{r^2}{\omega_0^2}\right) & \text{for } s \leqslant r \leqslant a \\ 0 & \text{otherwise} \end{cases} \quad (4.31)$$

where L_0 is the zeroth Laguerre polynomial as defined in appendix 1, a is the radius of the primary aperture and s is the radius of the secondary (centre stop).

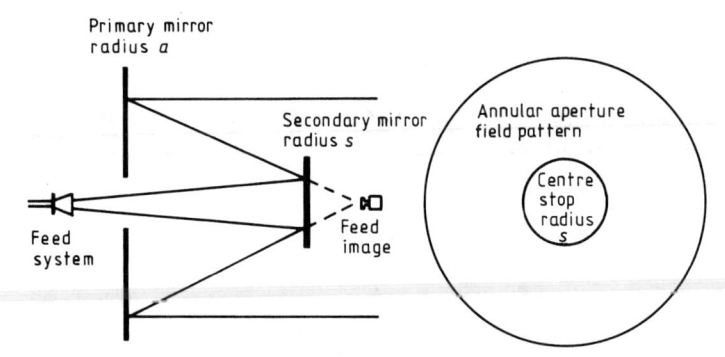

Figure 4.6 Simple representation of a Cassegrain antenna system. The mirror–feed combination produces an essentially annular pattern.

For the zeroth polynomial, $L_0 = \sqrt{2/\pi}$; hence the field within the unstopped aperture is

$$\Psi = \dfrac{1}{\omega_0} \sqrt{\dfrac{2}{\pi}} \exp\left(-\dfrac{r^2}{\omega_0^2}\right) \quad \text{for } s \leqslant r \leqslant a. \quad (4.32)$$

The power-coupling efficiency with which the field may be coupled through the finite stopped aperture will therefore be

$$N(a, s) = |\{\Psi | \Psi\}|^2 \quad (4.33)$$

where $\{\Psi | \Psi\}$ represents the coupling integral over the area bounded by the conditions $s \leqslant r \leqslant a$. For a fundamental Gaussian field the result of this integral is

$$N(a, s) = \exp\left(-\dfrac{2s^2}{\omega_0^2}\right) - \exp\left(-\dfrac{2a^2}{\omega_0^2}\right). \quad (4.34)$$

The value represents the efficiency with which the aperture is exploited by the feed and is therefore often referred to as an *aperture efficiency*.

The beam pattern which corresponds to the aperture field will be of the form

$$\phi(z) = \sum_p C_p \psi_p \qquad (4.35)$$

where ψ_p is the pth Gauss–Laguerre mode:

$$\psi_p = \frac{1}{\omega} L_p\left(\frac{2r^2}{\omega^2}\right) \exp\left[-j(kz - 2\pi ft) - j\Phi - r^2\left(\frac{1}{\omega^2} + \frac{jk}{2R}\right)\right]. \qquad (4.36)$$

The coefficients C_p may be defined from the coupling integral

$$C_p = \{\Psi \mid \psi_p\}. \qquad (4.37)$$

Since the aperture field is assumed to have a uniform phase distribution, it is sensible to assume that the ψ_p mode set has its beam waist located at the aperture plane. For simplicity we may also choose a beam waist size equal to ω_0. Hence we can write

$$C_p = \int_s^a \frac{1}{\omega_0^2} \sqrt{\frac{2}{\pi}} L_p\left(\frac{2r^2}{\omega_0^2}\right) \exp\left(-\frac{2r^2}{\omega_0^2}\right) 2\pi r \, dr. \qquad (4.38)$$

If a suitable book of standard integrals and the properties of polynomials is consulted, this can be shown to be equivalent to

$$C_p = \sum_{m=0}^{p} \frac{(-1)^m p!}{(m!)^2 (p-m)!} \int_S^A X^m \exp(-X) \, dX \qquad (4.39)$$

where

$$A \equiv \frac{a^2}{2\omega_0^2} \qquad S \equiv \frac{s^2}{2\omega_0^2}. \qquad (4.40)$$

The solution of this integral leads to the result

$$C_p = \sum_{m=0}^{p} \frac{(-1)^m p!}{(m!)^2 (p-m)!} \zeta \qquad (4.41)$$

where

$$\zeta \equiv m!\left[\left(1 + S + \frac{S^2}{2!} + \cdots + \frac{S^m}{m!}\right) \exp(-S)\right.$$
$$\left. - \left(1 + A + \frac{A^2}{2!} + \cdots + \frac{A^m}{m!}\right) \exp(-A)\right]. \qquad (4.42)$$

Using the above expressions we can now evaluate the coefficients C_p and hence define the field amplitude and phase at any point along the beam produced by the Cassegrain system.

In many cases we are mainly interested in the antenna's far-field pattern, i.e. the field when z approaches infinity. This field pattern will be of the form

$$\phi(\theta) = C_p L_p\left(\frac{2\theta^2}{\theta_0^2}\right)(-1)^p \exp\left(-\frac{\theta^2}{\theta_0^2}\right) \qquad (4.43)$$

where θ_0 is the beam divergence angle as defined by equation (4.26). The term $(-1)^p$ appears in this expression because the pth mode undergoes an anomalous phase shift of $p\pi$ between the beam waist ($z=0$) and far field ($z=\infty$). The above expression defines the field at an angle θ to the beam axis and is normalized such that an untruncated and unstopped fundamental Gaussian mode ($s=0$, $a=\infty$) whose waist size is ω_0 would contain a total beam power of unity.

Figure 4.7 shows a typical far-field pattern, calculated using the modes $m = 0, 1, \ldots, 10$. This illustration represents the case when $a/\omega_0 = 1.5$ and $s/\omega_0 = 0.2$. The broken curve represents the far-field pattern which would be produced by a fundamental Gaussian beam whose waist size was ω_0 and which contained a total power of $N(a,s)$. The full curve represents the Cassegrain system's far-field pattern, normalized to the on-axis ($\theta = 0$) level of the fundamental Gaussian shown. (That is, if the aperture efficiency is to be taken into account, the patterns should be multiplied by $N(a,s)$ to obtain the level at any angle θ/θ_0.)

Although we are generally interested in a far-field pattern, it is worth noting that this mode of Gaussian beam mode analysis provides us with a simple analytic technique for calculating the field at any plane in the near-field, midfield or far-field regions. Indeed the same approach can also be used to calculate the field at *any* plane along the beam, once the field is defined at just one plane. This can sometimes prove useful for, say, evaluating the aperture field from a measurement of the far-field (or midfield) pattern produced by an antenna.

The secondary mirror of a Cassegrain antenna reflects signal power back towards the antenna feed. If we consider the situation where the antenna is being employed to transmit a beam, some of the power radiated by the feed will pass back through the centre opening in the primary mirror and re-enters the feed. As a consequence, the Cassegrain system will appear as an unmatched load to the transmission line which supplies power to the feed.

This process can be analysed by thinking of the feed as being able to observe an image of itself in the secondary mirror. The feed beam pattern at the secondary mirror can be represented by a beam of field ψ and the reflected field—which appears to come from the feed's image—can be represented by a beam of field ϕ. For a circularly symmetric secondary mirror of diameter s, the field coupling between these two beams will be

$$C_s = \{\psi | \phi\}. \qquad (4.44)$$

In this case, $\{\psi | \phi\}$ represents a coupling integral over a finite-aperture plane defined by the secondary mirror's location and size. Taking the simplest case, where ψ and hence ϕ are fundamental Gaussian modes, we can therefore say that, for a secondary mirror of radius s,

$$C_s = \int_0^s \left(\frac{1}{\omega}\right)^2 \frac{2}{\pi} \exp\left[-r^2\left(\frac{2}{\omega^2} + \frac{jk}{2R} - \frac{jk}{2R'}\right)\right] 2\pi r \, dr \quad (4.45)$$

where $k = 2\pi/\lambda$, R and R' are the phase-front radii of the input and reflected beams and ω is the beam size at the secondary-mirror plane.

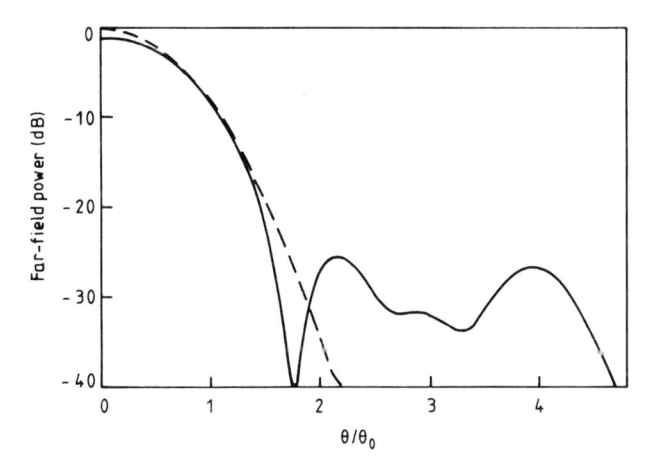

Figure 4.7 Calculated far-field pattern of a typical Cassegrain system. The power and angle are normalized as described in the text, $a/\omega_0 = 1.5$; $s/\omega_0 = 0.2$; aperture efficiency $= 0.91$, axial gain $= -1.03$ dB; 1/e field at angle $= 1.05$. A fundamental Gaussian beam curve (– – –) is shown for comparison.

This integral may be rearranged into the form of a standard complex integral, leading to the result

$$C_s = 2 \frac{1 - \exp[-s^2(\alpha + j\beta)]}{\omega^2(\alpha + j\beta)} \quad (4.46)$$

where

$$\alpha = \frac{2}{\omega^2} \qquad \beta = \frac{k}{2}\left(\frac{1}{R} - \frac{1}{R'}\right). \quad (4.47)$$

Knowing the radiation wavelength λ, the secondary radius s and the beam parameters ω, R and R', we can therefore calculate the effective reflectivity of a Cassegrain system produced by reflection at the secondary mirror.

Although the above example considers the antenna–feed as a transmitting system, this reflectivity is also important in receiving systems. The feed itself (or some other items in the receiver system) may also be mismatched, giving rise to a second partial signal reflection.

Consider for example the situation where the feed employed has a signal reflectivity ρ and the feed is at a distance Z from the secondary mirror. This system now behaves as if it were a resonator, formed from two semi-reflectors whose reflectivities are C_s and ρ, spaced a distance L apart. The total signal power lost by reflection will now vary periodically as a function of λ/L. In most practical cases, $L \gg \lambda$; hence quite small changes in signal wavelength may produce significant changes in signal coupling. For this reason, Cassegrain reflectivity problems may give rise to an unwanted frequency sensitivity in receiving or transmitting systems.

Chapter 5

Transmission Lines, Impedance Matching and Signal Reflections

5.1 Transmission Lines and Terminating Loads

A rectangular waveguide, an E-plane guide and even Gaussian beams in free space may all be considered as particular types of *transmission line*. A transmission line will have a uniform or periodic structure which guides electromagnetic waves in a given direction. The details of the structure vary from type to type—metal pipes, dielectric fibres, chains of lenses or mirrors, etc—but they all share a set of basic properties. For example, the field patterns in each case may be treated as *modes* propagating in a particular direction.

When dealing with very low signal frequencies, electronic systems can often be analysed by assuming that voltages and currents instantly move along connecting wires. At high frequencies, however, we often need to communicate signals between points which are many wavelengths apart, i.e. the time taken for signals to be transferred is long compared with the period of a cycle of the signal frequency. Under these conditions we can understand what happens by regarding the signals as being communicated via a transmission line.

Figure 5.1 illustrates a typical sort of transmission line called a *twin feeder* (or *Lecher line*). It consists of two parallel metal wires.

The basic properties of the arrangement can be seen by considering what happens if we attach an oscillator to one end of a very long piece of twin feeder. The oscillator produces a voltage

$$V(t) = V_0 \exp(j2\pi ft) \tag{5.1}$$

between the wires at the place where it is attached.

The applied voltage produces a potential difference between the wires; hence there is an electric field between them. As the wires have a finite surface area, this means that they must become charged, i.e. the generator must produce a current flow from one wire into the other, to produce the electric field. Hence the generator also produces a magnetic field around the wires.

The electromagnetic wave flows away from the generator, guided by the electrons which move around in the surfaces of the two wires. If the twin feeder is *very* long (i.e. essentially infinite) the wave flows away and is never seen again by the generator. Signal power flows away from the generator as if the generator were connected to a resistive load. The capacitance and inductance per unit length of the wires—and hence the effective resistance of a very long run of twin feeder—will depend upon their diameters and spacing.

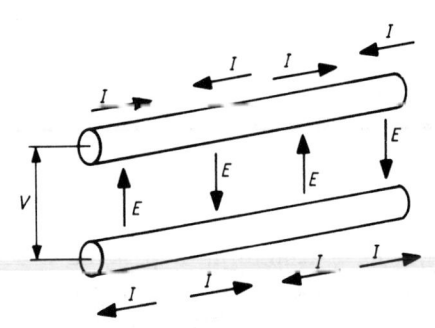

Figure 5.1 'Twin feeder' transmitting an electromagnetic signal.

The voltage between the wires at some place a distance x from the generator will be

$$V(t, x) = V_0 \exp\left[j2\pi\left(ft - \frac{x}{\lambda}\right)\right] \quad (5.2)$$

where λ is the wavelength of the signal whose frequency is f.

The current $I(t, x)$ on the wires will be proportional to the voltage. We can therefore say that

$$I(t, x) = \frac{V(t, x)}{Z_c} \quad (5.3)$$

where the value Z_c depends upon wire sizes and separation. Z_c is therefore often called the *characteristic* or *waveguide impedance* of the transmission line.

If we assume that the metal wires have a negligible resistivity, the basic properties of the transmission line can be described in terms of just two parameters:

(i) the value of the *characteristic impedance* Z_c;
(ii) the signal velocity $c = \lambda f$.

For many purposes we do not need to know the details of the field patterns which propagate along the line. So far as the generator is concerned, an infinite length of line looks the same as a resistor, whose resistance R equals Z_c. A twin feeder and a number of other forms of transmission line, including free-space beams and *coaxial cables*, share the ability to transmit TEM waves which are not cut off at a specific low frequency. Other forms of transmission line, e.g. a rectangular waveguide, do not have this property. Despite this, all these systems can be treated adequately as transmission lines for most purposes.

From the assumption that the wires themselves do not dissipate any signal power it follows that any signal put in by the generator moves along the transmission line without altering in any way. The signal which arrives at the 'output' end of a long line is identical with that which was put into the 'input' end by the generator some time ago. As a result, if we attach a resistance $R = Z_c$ to the other end of the line, it sees a 'delayed' version of the signal produced by the generator. (For this reason, transmission lines are sometimes used as 'delay lines' to produce time delayed versions of a signal.)

Unless the resistor is provided with some sort of clock, it has no way of knowing how far away it is from the generator. The voltage between the wire ends is V and, obeying Ohm's law, the resistor passes a current $I = V/R = V/Z_c$. This means that the generator cannot tell the difference between an infinite line of impedance Z_c and a shorter line terminated in a resistance $R = Z_c$.

A resistor with $R = Z_c$ is said to be *matched* to the transmission line. All the signal power transmitted along the line from the generator is coupled perfectly into such a resistor. If, however, the resistor is replaced by an impedance Z_1 which does *not* equal Z_c, then the arrangement is said to be *mismatched*.

Consider the example of a termination $Z_1 = 0$, i.e. the line is terminated in a short circuit. The wave propagating along the line will produce a voltage $V(t, x)$ and current $I(t, x)$. However, we cannot produce any voltage across a short circuit. Hence the voltage at the termination must always be zero, no matter how large the current.

The wave will carry a mean signal power P of $\text{Re}(\frac{1}{2} VI^*)$. None of this power can be dissipated in the short circuit as this would require us to produce a non-zero voltage across zero impedance. The power cannot just vanish quietly; hence it takes the only way out and flows back along the line. All the signal power coupled via a transmission line onto a short circuit is reflected back along the line.

The line now carries *two* waves, propagating in opposite directions. The voltage and current that we measure at any place along the line is composed of the sum of these two waves.

More generally, when a signal propagating along a transmission line is incident upon a termination, a fraction of the signal power is absorbed by the termination and the rest is reflected back along the line. The fractional power *reflectivity* will depend upon how the termination impedance Z_1 compares with the characteristic impedance Z_c.

An incident signal with

$$V_i = V_0 \exp\left[j2\pi\left(ft - \frac{x}{\lambda}\right)\right] \tag{5.4}$$

will generate a reflected wave

$$V_r = \rho V_0 \exp\left[j2\pi\left(ft + \frac{x}{\lambda}\right)\right] \tag{5.5}$$

where ρ is a *reflection coefficient* which determines the relative magnitude and phase of the reflected wave. Each wave will have an associated current

$$I_i = \frac{V_i}{Z_c} \tag{5.6}$$

and

$$I_r = -\frac{V_r}{Z_c}. \tag{5.7}$$

The difference between the signs of x/λ in equations (5.4) and (5.5) indicates that the waves are moving in opposite directions. The opposite signs in equations (5.6) and (5.7) may be thought of as a consequence of the requirement that one wave carries power *to* the termination and the other carries it *away* from the termination.

The voltage V_1 across and the current I_1 through a terminating impedance Z_1 placed a distance $x = d$ away from the generator will hence be

$$V_1 = V_i + V_r \tag{5.8}$$

and

$$I_1 = I_i + I_r \tag{5.9}$$

where

$$V_1 = Z_1 I_1. \tag{5.10}$$

Rearranging these expressions, we obtain the result

$$\rho = \exp(-2jd\beta)\frac{Z_1 - Z_c}{Z_1 + Z_c} \tag{5.11}$$

where β is the *propagation constant*:

$$\beta = \frac{2\pi}{\lambda}. \tag{5.12}$$

From equations (5.4) and (5.5) it can be seen that, for $x = t = 0$, ρ gives the relative magnitude and phase of V_r compared with V_i, i.e. ρ is the effective reflectivity of the load–transmission line combination as seen by the generator. The actual load reflectivity would not change if a different line length d were chosen. Hence it is useful also to define the *load* reflectivity Γ, which is equal to ρ if $d = 0$, i.e. we may write

$$\Gamma = \frac{Z_1 - Z_c}{Z_1 + Z_c} \qquad (5.13)$$

and

$$\rho = \Gamma \exp(-2jd\beta). \qquad (5.14)$$

A matched load corresponds to the condition $Z_1 = Z_c$ and produces the results $\Gamma = 0$ and $\rho = 0$, i.e. no power is reflected back towards the generator. We have already observed that terminating the line in a short causes all the power to be reflected. The *power* reflectivity will vary as $|\Gamma|^2$. For a short ($Z_1 = 0$) the load reflectivity $\Gamma = -1$, i.e. the reflected and incident waves are 180° out of phase.

If the termination impedance Z_1 is infinite (an 'open' circuit through which *no* current may flow), then $\Gamma = 1$ and all the input signal power is reflected. In this case the incident and reflected signals are in phase at the terminal load.

The effect of a given termination load and length of transmission line is often treated as being equivalent to an appropriate impedance connected directly to the generator terminals. The effective impedance Z seen by the generator can be obtained from

$$Z = \frac{V}{I} \qquad (5.15)$$

where

$$V = V_0 \exp(2\pi j ft) + \rho V_0 \exp(2\pi j ft) \qquad (5.16)$$

$$I = \frac{V_0}{Z_c} \exp(2\pi j ft) - \frac{V_0}{Z_c} \rho \exp(2\pi j ft). \qquad (5.17)$$

Hence we can show that

$$Z = Z_c \frac{1+\rho}{1-\rho}. \qquad (5.18)$$

A case which is of particular interest is the use of a short circuit ($\Gamma = -1$) which may be moved backwards and forwards along the transmission line. The arrangement is generally referred to as a *back-short*. From equation (5.14) the reflection coefficient ρ produced by a short will be

$$\rho = -\exp(-2j\beta d) \qquad (5.19)$$

and the effective load impedance presented to the generator will be

$$Z = jZ_c \tan(\beta d). \qquad (5.20)$$

This result has two important features. Firstly, it shows that the effective impedance produced by a back-short is purely *reactive*. Secondly, it indicates that we may produce any effective load reactance from $+j\infty$ to $-j\infty$ by choosing an appropriate value for the distance d.

Thus far we have only considered a load which terminates a length of transmission line. In practice, many devices act as a load placed across a line at some place other than the end. Such a load behaves like a 'shunt', allowing some current to flow between the wires and producing a reflected wave. However, unlike a termination, some signal power may be transmitted past the load along the remaining length of transmission line. A similar situation arises if one length of transmission line is coupled to another which has a different characteristic impedance. In either case, some power will be reflected and some transmitted.

Figure 5.2(*a*) illustrates an 'input' length of transmission line of characteristic impedance Z_c connected to a load Z_1 via an intermediate length d of line whose characteristic impedance is Z_c'. A signal may be injected into the system by connecting a generator to the input end of Z_c.

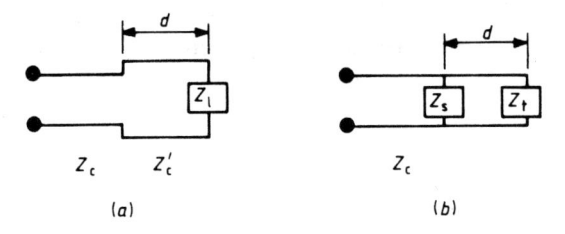

Figure 5.2 (a) Transmission line coupled to a length d of mismatched line and a terminating load Z_1. (b) Transmission line with a shunt load Z_s located a distance d from the terminating load Z_1.

The combination of the terminating impedance and the interconnecting line will behave as if an effective impedance Z were connected to the end of the input line. When equations (5.11) and (5.18) are combined the effective impedance terminating the input line will be

$$Z = Z_c' \frac{1 + \Gamma \exp(-2jd\beta)}{1 - \Gamma \exp(-2jd\beta)} \qquad (5.21)$$

where

$$\Gamma = \frac{Z_1 - Z_c'}{Z_1 + Z_c'}. \qquad (5.22)$$

If we can alter the length d, it is sometimes possible to arrange that $Z = Z_c$ even though $Z_c \neq Z_c'$ and $Z_1 \neq Z_c'$. In such cases the intermediate line acts as a *transformer* which matches the terminal load to the input transmission line. All the input signal power is then coupled into the load without any reflection loss.

Figure 5.2(b) illustrates a length of line which is terminated in an impedance Z_t but which also has a shunt impedance Z_s placed across it at a distance d from the termination. In general, both Z_t and Z_s will generate reflected waves but—as in the case of a terminated length of a different line—it is sometimes possible to arrange that the *total* reflectivity is zero.

At the place where Z_s is connected, the combination of the termination Z_t and the line of length d will produce an effective impedance Z_e where

$$Z_e = Z_c \frac{1 + \Gamma \exp(-2jd\beta)}{1 - \Gamma \exp(-2jd\beta)} \tag{5.23}$$

where

$$\Gamma = \frac{Z_t - Z_c}{Z_t + Z_c}. \tag{5.24}$$

The total impedance Z which appears to be placed at this position will be such that

$$\frac{1}{Z} = \frac{1}{Z_s} + \frac{1}{Z_e}. \tag{5.25}$$

The overall reflectivity of the system will be zero if $Z = Z_c$, i.e. if we arrange that

$$\frac{1}{Z_c} = \frac{1}{Z_s} + \frac{1}{Z_e}. \tag{5.26}$$

This example differs slightly from the previous one in that the input power may be shared between Z_t and Z_s. In both cases, however, the system has been matched to the input guide and no power is lost by reflection. In many practical systems the arrangements discussed above are used to ensure efficient signal coupling into a device. For coherent signals these arguments are also valid when the wave direction is reversed. Hence power coupling *out* of a source of a given impedance may also be optimized in an identical manner.

One of the simplest, and most widely used, methods for matching power into (or out of) a device is to place the device as a shunt at a given distance from a short which terminates a transmission line. For a device of impedance $Z = X + jY$ the ideal line impedance is $Z_c = X \times [1 + (Y^4 + X^2Y^2)/(X^4 + X^2Y^2)]$. The short is then placed at a distance d away from the device such that

$$X^2/Y + Y = Z_c \tan(\beta d). \tag{5.27}$$

The combination of the short and the device is then matched perfectly and no signal power is reflected.

This simple arrangement does have some drawbacks; for example, the choice of a given short distance d 'tunes' the arrangement so that perfect matching only occurs at specific β values. If the signal frequency alters significantly, it becomes necessary to move the short to rematch signal power into the device. Problems can also arise for some devices where the real part X of the device impedance is very large or very small. This makes it difficult to produce a satisfactory transmission line with the required characteristic impedance.

5.2 Semi-reflectors, Dielectric Slabs and Mismatched Lines

More complex arrangements of shunts, changes in line impedance, and back-shorts may be used to obtain improvements in particular cases. By choosing an arrangement which has a *high* signal reflectivity at a signal frequency it becomes possible to make *filters* which prevent unwanted signals from being coupled into a device. Hence we can make systems which act as matching transformers and signal filters. In each case the properties of the system can be examined in terms of their signal reflection, transmission and absorption effects.

Free-space devices, acting upon Gaussian beams, often behave as *semi-reflectors*. Various types of device exist and these will be considered in more detail in later chapters. These devices may also be treated as shunt or termination impedances coupled into a transmission line which represents the free-space beam.

When dealing with low signal frequencies, we can often determine impedances by making voltage and current measurements. At millimetre-wave frequencies, and for free-space beams, this is impractical. Instead, the behaviour of a device is determined by measuring its reflection, transmission and absorption properties.

Consider a device whose impedance is Z_s which is placed as a shunt across a line of characteristic impedance Z_c. This may be taken to represent a device placed in a Gaussian beam which will reflect, transmit and absorb a given amount of the incident signal power.

So far as the signal incident upon the device is concerned, the combination of the device and the continuing line will behave as if they were a *termination* whose impedance Z_1 is such that

$$\frac{1}{Z_1} = \frac{1}{Z_c} + \frac{1}{Z_s}. \tag{5.28}$$

Transmission Lines, Impedance Matching and Signal Reflections

In free space the impedance is defined in terms of the ratio of the electric and magnetic fields. Here we can regard the signal 'voltage' as related to the magnitude of the beam electric field distribution.

The device's field reflectivity will be

$$\Gamma = \frac{Z_1 - Z_c}{Z_1 + Z_c}. \tag{5.29}$$

If the input signal incident upon the device is

$$V_i = V_0 \exp(2\pi j f t) \tag{5.30}$$

then the total signal magnitude seen by the (apparently terminating) load will be

$$V = V_i + V_r \tag{5.31}$$

where the reflected signal V_r is

$$V_r = \Gamma V_i. \tag{5.32}$$

Some signal power will usually be transmitted past the device and, viewed from the 'output' side of the line, the device will appear as a source which transmits a signal $V_t = V$ along the rest of the line, i.e. we can say that

$$V_t = V_i + \Gamma V_i = \left(1 + \frac{1}{\Gamma}\right) V_r. \tag{5.33}$$

Hence the impedance of the device can be obtained by measuring the reflected and transmitted signals produced by a known input.

Many practical devices act as semi-reflectors without absorbing significant amounts of signal power. These devices can be regarded as purely reactive loads (i.e. having no real resistance). For such a device the impedance can be written as

$$Z_s = jX. \tag{5.34}$$

This produces an apparent termination impedance as seen by the incident signal of

$$Z_1 = \frac{jXZ_c}{jX + Z_c}. \tag{5.35}$$

Hence

$$\Gamma = -\frac{Z_c}{2jX + Z_c} \tag{5.36}$$

from which it follows that the transmitted and reflected fields are related by the expression

$$V_t = -\left(\frac{2jX}{Z_c}\right)V_r. \qquad (5.37)$$

From this expression it can be seen that, for any non-absorbing semi-reflecting device, the transmitted and reflected fields are 90° out of phase at the device. This result is general. It does not depend upon the nature of the device or the transmission line. It also has important consequences in arrangements which employ a number of semi-reflecting elements to achieve specific effects (e.g. in Fabry–Perot filters).

This result may also be seen as a consequence of the principle of energy conservation and the requirement that the field must be continuous and single valued. The input and output fields must be such that

$$V_i + V_r = V_t. \qquad (5.38)$$

However, the principle of energy conservation requires that

$$|V_i|^2 = |V_r|^2 + |V_t|^2. \qquad (5.39)$$

Unless the fields are all zero these two expressions can only both be true if we allow the voltages to be vectors and arrange that V_r and V_t are orthogonal. In the arguments presented above, the vector nature of these fields show itself as each field having real and imaginary components and V_t and V_r are at 90° (i.e. orthogonal) in time.

Many of the systems and devices considered in later chapters reflect and transmit signals by exploiting polarization effects. In these cases the reflected and transmitted signals may be orthogonally polarized in space (e.g. plane waves polarized at right angles to each other). The fields can then be in phase in time whilst continuing to satisfy the above requirements. However, for any loss-free semi-reflector, it remains true that the reflected and transmitted signal must be orthogonal—either in time or in space.

For a wave propagating in free space the ratio of the electric to magnetic fields has the value 377 Ω. This resistance is the *impedance of free space* and may be used for Z_c when a free-space beam is treated as a wave propagating along a transmission line. A wave moving inside a dielectric medium may have a E/H field ratio which differs from 377 Ω. This change in E/H is described in general optics in terms of an appropriate *relative permittivity* ε_r (or a refractive index μ).

When a beam moving through a dielectric is considered, it is convenient to regard the beam as propagating along a transmission line whose characteristic impedance differs from that of free space. The signal reflection which occurs at a dielectric interface can then be treated as arising because the surface behaves as a junction between two transmission lines of different characteristic impedances.

The signal reflectivity Γ of a junction between two lines of differing impedances Z_c and Z_c' is

$$\Gamma = \frac{Z_c' - Z_c}{Z_c' + Z_c} \tag{5.40}$$

(for a signal incident along the Z_c line).

As both Z_c and Z_c' are real, Γ is also real, i.e. the reflected signal is either in phase or 180° out of phase with the incident wave at the junction.

If V_i represents the incident field, then the total field at the junction will be

$$V_t = V_i(1 + \Gamma). \tag{5.41}$$

The field V_t acts as a source transmitting a wave along the line whose impedance is Z_c', i.e. V_t is the transmitting signal at the junction. If we define the *transmission coefficient* τ such that

$$V_t = \tau V_i \tag{5.42}$$

then we can write that

$$\tau = 1 + \Gamma. \tag{5.43}$$

The incident signal power is V_i^2/Z_c and the reflected signal power is $(\Gamma V_i)^2/Z_c$. As no power is absorbed at the junction, we can say that

$$\frac{\tau^2}{Z_c'} = \frac{1}{Z_c} - \frac{\Gamma^2}{Z_c}. \tag{5.44}$$

Equations (5.43) and (5.44) both lead to the result

$$\tau = \frac{2Z_c'}{Z_c + Z_c'}. \tag{5.45}$$

As Γ and τ are real, the transmitted signal V_t and the reflected signal V_r must both be either in phase or 180° out of phase with the incident signal V_i at the junction.

The reflection properties of a dielectric interface and of a parallel-sided slab have already been discussed in chapter 2. There it was pointed out that the field reflectivity at an interface where an incident signal passes from a medium of refractive index μ_1 to another of refractive index μ_2 will be

$$\frac{\mu_2 - \mu_1}{\mu_1 + \mu_2}. \tag{5.46}$$

Hence we can identify the refractive index of a low-loss dielectric as being equivalent to the characteristic impedance of the medium when considered as a transmission line. Each surface of a dielectric slab can be treated as a junction between two transmission lines of differing impedances. The overall power transmission and reflection can then be calculated using a transmission line approach.

In general, systems involving a series of semi-reflecting and dielectric elements may also be analysed using transmission line arguments to calculate the effective load (and hence reflectivity) presented to the input signal. These methods can be used for both conventional waveguide and free-space beam systems to design a wide range of instruments and devices. Some examples of this technique will be mentioned in later chapters.

Chapter 6

Bolometric Signal Detection

6.1 Basic Properties of Thermal Detectors

Many different sorts of millimetre-wave detectors have been developed, each offering its own combination of advantages and drawbacks. For convenience, they may be divided into two general categories: bolometers (heat detectors), and mixers (heterodyne detectors). In this chapter we shall consider bolometric signal detection systems.

A bolometer is a device which responds to a change in temperature produced when it absorbs incident radiation. Most bolometers are relatively unfussy regarding the frequency of incident radiation. Incident power of whatever signal frequency is absorbed, the element temperature changes, and this temperature change is sensed. Bolometers are therefore particularly useful when we are mainly concerned with measuring the input signal power. If we need to determine the power or frequency spectrum of the incoming signals, we can employ a suitable filter which allows only given signal frequencies to fall upon the bolometer. Systems such as the *polarizing interferometer* discussed in a later chapter may be regarded as a form of variable filter which permit us to explore the power or frequency spectrum of a signal without having to replace one filter with another.

The behaviour of a bolometer is basically dependent upon its thermal properties. In most cases any change in bolometer temperature is sensed electronically. For example, a bolometer may have an electrical resistance which varies with temperature. Consider the idealized bolometer illustrated in figure 6.1.

A typical system consists of a heat-sensitive detector element mounted inside a heat sink which almost surrounds the detector. The input signal power may be coupled onto the detector through an opening in the immediate surroundings. A 'window' is usually placed over the opening. This can be used to prevent power at unwanted signal frequencies from reaching the detector. The detector element is physically supported by an arrangement designed to provide a controlled heat conduction path between it and the surrounding heat sink. Wires will also be connected to the element so that electronic changes produced by input signals may be sensed.

The bolometer is assumed to have a thermal capacity C and is connected to a heat sink of temperature T_0 via a thermal conductivity s. For a real bolometer, heat may be coupled in or out either by conduction through its physical supports or by radiation exchanging power with its surroundings. Hence the value of s will depend upon both conduction and radiation.

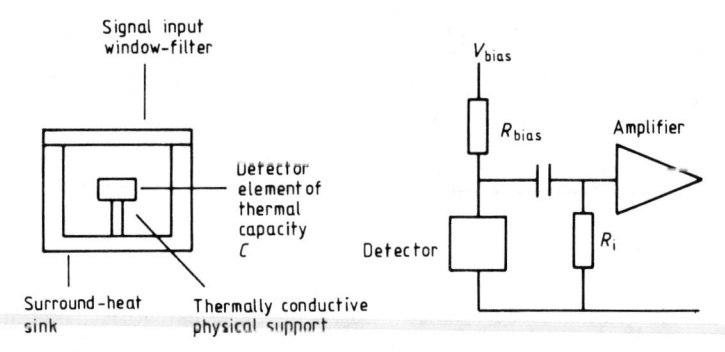

Figure 6.1 Thermal (left) and electronic (right) models of a bolometric detector.

The signal power P absorbed by the detector element and its temperature T will be related by the expression

$$P = C\frac{dT}{dt} + s(T - T_0). \qquad (6.1)$$

If the input signal power is fixed at a particular level $P = F_0$, then the element temperature will change exponentially towards the value

$$T = \frac{F_0}{s} + T_0. \qquad (6.2)$$

In most applications we are concerned with how the signal power varies with time. For an input signal power which varies as

$$P = F_0 + \delta F \sin(\omega t) \qquad (6.3)$$

the detector element temperature will be

$$T(t) = T' + \delta T \sin(\omega t - \phi) \qquad (6.4)$$

at the time t where

$$T' = \frac{F_0}{s} + T_0 \qquad (6.5)$$

is the mean element temperature,

$$\delta T = \frac{\delta F}{\sqrt{C^2\omega^2 + s^2}} \qquad (6.6)$$

is the magnitude of the element's temperature variation, and

$$\phi = \arctan\left(\frac{C\omega}{s}\right) \qquad (6.7)$$

is the phase lag between the input power fluctuation and the temperature variation that it produces.

Any change δT in the detector temperature will produce a change in the electrical properties of the device. In figure 6.1 it is assumed that the electrical resistance of the bolometer changes with its temperature. A *bias current* may be passed through the bolometer so that the change in resistance will produce a change δV in the voltage across the element.

If a given change δF in input signal power produces a voltage change δV, then the sensitivity of the bolometer used as a detector can be defined in terms of the *responsivity* $\mathcal{R}$:

$$\mathcal{R} = \frac{\delta V}{\delta F} \qquad (6.8)$$

(In some cases the responsivity will be defined in terms of a *current* change rather than a voltage change.)

The responsivity will be proportional to $\delta T/\mathrm{d}F$; hence from equation (6.6) we can say that

$$\mathcal{R} \propto \frac{1}{\sqrt{C^2\omega^2 + s^2}}. \qquad (6.9)$$

For $\omega \ll s/C$

$$\mathcal{R} \propto \frac{1}{s} \qquad (6.10)$$

i.e. at relatively low modulation frequencies the bolometer's responsivity essentially depends upon only the effective thermal conductivity s.

For $\omega \gg s/C$,

$$\mathcal{R} \propto \frac{1}{C\omega} \qquad (6.11)$$

i.e. at relatively high modulation frequencies the responsivity depends upon the bolometer's thermal capacity and falls with increasing modulation frequency.

We can define the bolometer *response time* or *time constant* τ to be

$$\tau = \frac{C}{s} \qquad (6.12)$$

and rewrite equation (6.9) as

$$\mathscr{R} \propto \frac{1}{s\sqrt{1+\omega^2\tau^2}}. \tag{6.13}$$

The phase lag ϕ between the input signal modulation and the output voltage variation can also be rewritten as

$$\phi = \arctan(\tau\omega). \tag{6.14}$$

τ indicates the time required for the detector to respond to a change in the input signal power level. In general, we would like to produce a bolometer which has a high responsivity. One possible way to increase the bolometer responsivity is to reduce the conductivity s. However, this has the effect of increasing τ, which means the detector will take longer to respond to changes in the signal power.

Reducing the thermal capacity C will improve the response time. Although this has no effect upon the responsivity at a low modulation frequency, it is a useful way of preventing the response time from becoming too long when the thermal conductivity is reduced. Unfortunately, reductions in thermal capacity generally mean that the detector element has to be physically smaller. However, if the detector is too small, it may become difficult to couple the input signal power efficiently.

Thermal energy may usually flow in and out of a bolometer element in two ways: by conduction and by radiation. The conductivity s should therefore really be considered as the sum of two conductivities. For a constant input signal power level P the bolometer temperature T will therefore be such that

$$P = (T - T_0)G + A\sigma_S(T^4 - T_0^4) \tag{6.15}$$

where G is the conductivity for thermal conduction through the detector element's physical support, A is the detector's surface area and σ_S is Stefan's constant. From equation (6.15) we can say that

$$\frac{dT}{dP} = (G + 4\sigma_S A T_0^3)^{-1} \tag{6.16}$$

i.e. the total thermal conductivity s will be

$$s = G + 4\sigma_S A T_0^3 \tag{6.17}$$

provided that the input signal power is small and T is almost the same as the temperature T_0 of the heat sink and surroundings of the bolometer element.

Equation (6.17) leads us to two conclusions. Firstly, no matter how small we make the physical conductivity G, the total thermal conductivity cannot be less than $4\sigma_S A T_0^3$. Secondly, when G is small, the overall conductivity (and hence the detector responsivity) is strongly dependent upon the temperature of the detector and its surroundings.

Bolometric Signal Detection

Assuming that $G \simeq 0$ we can calculate that the ratio of the responsivity of a given detector at 3 K to its responsivity at 300 K will (if all else is equal) be

$$\frac{\mathcal{R}(3\text{ K})}{\mathcal{R}(300\text{ K})} = \frac{300^3}{3^3} = 1\,000\,000. \tag{6.18}$$

This is one of the main reasons why bolometric detectors are often cooled to low temperatures. In practice, G will not be zero and, at a suitably low temperature T_0, $G \simeq 4\sigma_S A T_0^3$. Reducing the temperature of the detector and its surroundings below this value will then not produce a significant further improvement in responsivity. It can be seen, however, that—unless the detector cannot work well at a reduced temperature—the responsivity can be dramatically improved by cooling.

Figure 6.2 shows a diagram of a typical *cryostat* system which enables a bolometer to work at low temperatures.

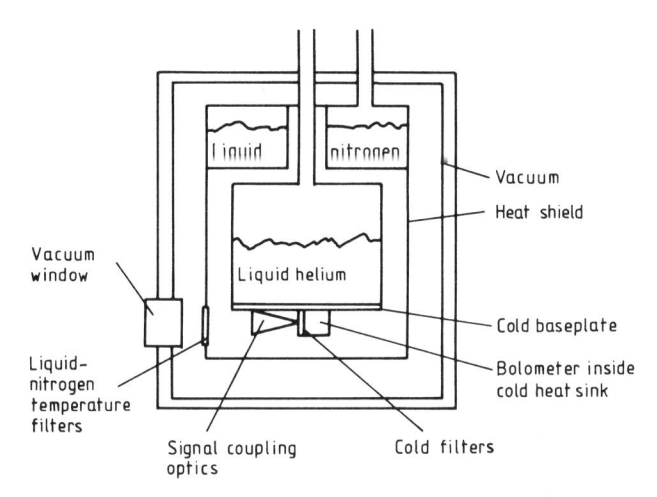

Figure 6.2 Typical liquid-helium cryostat system for cooling a low-temperature bolometric detector.

The detector is mounted inside a cavity or guide which is connected to the metal baseplate of a liquid-helium vessel. This will cool the detector and its immediate surroundings to a temperature of around 4 K. By using a vacuum pump it is possible to lower the vapour pressure above the liquid helium. In this way the detector temperature may be reduced still further.

As far as possible, unwanted thermal energy is kept away from the detector and its mount. Unwanted energy reaching the detector will tend to increase its temperature and to degrade its performance. Apart from a series of filters which act as 'windows' designed to pass only wanted signal energy, the detector is surrounded by *heat shielding*. The system illustrated in figure

6.2 shows a *nitrogen shield*—a surrounding screen of metal cooled to around 77 K by a liquid-nitrogen vessel.

Because of the low internal temperatures, air must be excluded to prevent freezing and condensation inside the vessel. The detector and its surroundings are therefore evacuated. One of the main functions of the outermost window is to keep air out of the system. The inner window—shown here in the nitrogen shield—is designed to stop radiation at unwanted frequencies from reaching the detector. A final 'cold' filter is often placed in the detector mount or its input optics to reduce still further the unwanted heat radiated onto the detector. These filters also prevent the detector from producing spurious output in response to signals at frequencies outside the range of interest.

Two general types of filter can be used: absorbing or reflective. An absorption filter will absorb heat radiation at unwanted signal frequencies. The emissivity and absorptivity of a material are always the same. Hence a filter which absorbs strongly in a given frequency range will also radiate strongly as a 'black body' over the same range. However, the filter is much colder than the thermal sources outside the system. Hence the net effect is to reduce greatly the unwanted thermal power seen by the detector.

A reflection filter reflects unwanted input power away from the detector. At frequencies where it is highly reflective it will not emit any significant power but, instead, reflects back any radiation that it receives from the detector or the inside of its mount. Hence the detector sees a cold reflection of itself in the filter.

So far we have discussed the responsivity (i.e. the sensitivity) of a bolometer. This can be regarded as a sort of 'conversion efficiency' which determines how large an electrical output will be produced by a given input signal. The ability of the bolometer to detect very small signals will ultimately be limited by the amount of *random noise* present in the system.

The presence of random noise is a general property of physical systems and is a consequence of the underlying statistical quantum mechanical nature of the real world. Electronic charge and electromagnetic radiation are both quantized. When dealing with small signals we must consider the effects of random fluctuations in the movements of electrons and photons upon any measurements that we wish to make.

Here we shall only consider the three main types of noise which arise in bolometer systems. Figure 6.3 illustrates the physical processes which generate these types of noise.

Figure 6.3(*a*) shows a piece of resistive material with a pair of electrical contacts attached to its ends. Random thermal motions of the electrons inside the material cause the internal charge distribution to fluctuate unpredictably from moment to moment. The potentials of the end contacts will depend upon the nearby charge distribution; hence there is a randomly varying potential difference between the two contacts.

Bolometric Signal Detection

If we connect a voltmeter to the two contacts, we observe a randomly varying voltage. The average voltage will be zero, being just as often negative as positive. The *squared* voltage will, however, never be negative; hence the amount of noise present may be measured in terms of the mean squared voltage. In practice, the amount of noise is often determined in terms of the average root mean square voltage or the mean noise *power*.

Any real measurement system will take a finite amount of time to respond to a change in voltage. Fluctuations which take place over a period shorter than this *response time* will therefore have no effect upon the measured voltage. The finite response time means that the measurement only takes into account noise fluctuations over a limited frequency range or *bandwidth*.

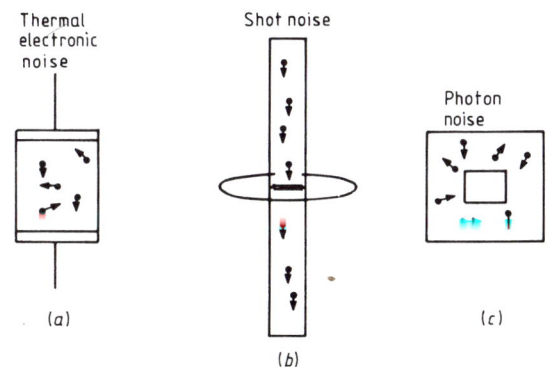

Figure 6.3 (a) Thermal motion of charge carriers in a resistor. (b) Flow of quantized charges along a conductor. (c) Energy exchange by photo emission and absorption.

The amount of noise generated by a resistance R at absolute temperature T may be represented in terms of a 'noise voltage generator' placed in series with the resistance which produces a mean squared voltage

$$e_n^2 = 4kTRB \tag{6.19}$$

where k is *Boltzmann's constant* ($k = 1.38 \times 10^{-23}$ Ws K^{-1}) and B is the bandwidth of the measurement system.

The noise power which this produces at the input of a measurement system will depend upon the system's input resistance R_i. The noise voltage e_n sets up a current i which must flow through R and R_i in series. Hence, from Ohm's law,

$$i = \frac{e_n}{R + R_i}. \tag{6.20}$$

This means that the *observed* noise voltage e_0 generated at the system input will be

$$e_0 = iR_i = e_n \frac{R_i}{R + R_i}. \quad (6.21)$$

The mean noise power N observed at the system input will therefore be

$$N = e_0 i = e_n^2 \frac{R_i}{(R + R_i)^2}. \quad (6.22)$$

This has a maximum value if $R = R_i$ when

$$N(\text{max}) = \frac{e_n^2}{4R}. \quad (6.23)$$

This value is generally referred to as the *maximum available noise power*.

Thermal noise is often called *Johnson* noise after its discoverer. It is sometimes also referred to as *Nyquist* noise in recognition of the person who provided the first theoretical analysis of its origin.

Another form of electronic noise arises when current flows along a conductor. Figure 6.3(*b*) represents a wire along which a current is flowing. A 'steady' current actually consists of a stream of discrete charge carriers passing along the wire. The average current I_0 flowing through a plane cutting the wire can therefore be written as

$$I_0 = \frac{qn}{t} \quad (6.24)$$

where q is the magnitude of each discrete charge and n is the mean number which cross the plane in the time interval t.

In reality the moving carriers will have a random distribution of individual velocities imposed on their overall drift velocity along the wire. The number crossing a specific plane will vary randomly from one short time period t to the next. The actual current will therefore vary randomly from instant to instant about the mean level I_0. Averaged over a particular time interval of length t, the measured current I will be

$$I = I_0 + i_s \quad (6.25)$$

where i_s is a random noise current whose mean squared size can be shown to be

$$i_s^2 = 2qI_0 B. \quad (6.26)$$

This type of noise is called *shot noise*. As with thermal noise the amount observed depends upon the measurement system bandwidth B. This bandwidth corresponds to the range of frequencies which will produce a response in a system which averages over the time period t. Unlike thermal noise, shot noise depends upon the mean current flow but not the physical temperature.

A number of other physical processes generate different types of noise, e.g. $1/f$ *noise* whose magnitude depends upon the physical construction of the electronic components. Thermal and shot noise, however, arise from the discrete nature of the fundamental charge carriers and the quantum mechanical nature of the real world. Hence these other forms of *excess noise* may—in principle at least—be alleviated by our skill in producing electronic devices. Thermal and shot noise cannot be reduced in this way. Instead they can only be dealt with by lowering the temperature T of a resistive element or reducing the mean current level I_0.

The third form of noise which is important in bolometers is *photon noise*. This is illustrated in figure 6.3(c). Electromagnetic radiation is also quantized. Power is radiated back and forth between the detector element and its environment, producing a radiative contribution to the bolometer's thermal conductivity. The quantized energies of the exchanged photons and the random statistics of the exchange mean that the net heat flow into or out of the bolometer varies randomly from instant to instant.

A similar effect also arises in the flow of heat by conduction through the physical support of the detector element. The conducted heat is also quantized into *phonons*. Their behaviour within the solid supports is broadly similar to that of the photons in free space.

The random fluctuations in the bolometer's temperature will depend upon the temperature of the bolometer and its surroundings. It will also depend upon the range of photon frequencies which can be absorbed and/or emitted by the bolometer and the signal power level (which is also a stream of photons). The signal power received will fluctuate from instant to instant owing to the random statistics of the photons which make up the signal. This effect is similar to electrical shot noise in that the noise level generated depends upon the signal level.

At very low signal power levels the photon noise becomes dominated by the thermal radiative exchange between the detector and its surroundings. It can be shown that, under these circumstances, the random fluctuations in the temperature of the bolometer element will be equivalent to a fluctuating signal power input δW whose mean squared size is

$$(\delta W)^2 = 4kT^2 GB \tag{6.27}$$

where T is the temperature of the bolometer and its surroundings and B is the measurement bandwidth.

The noise performance of a bolometer is often given in terms of a *noise equivalent power* (NEP). This is the input signal power which is equal to the mean apparent fluctuation in the input power produced by noise (usually defined for a standard bandwidth of 1 Hz).

For a 1 Hz bandwidth and for a detector where the thermal conductivity G is essentially radiative, the minimum NEP (ignoring electrical noise) will

be
$$\text{NEP} = 4\sqrt{A\sigma_S k T^5 B} \qquad (6.28)$$

where σ_S is Stefan's constant and A is the detector's absorbing area. For a detector whose area is 1 cm^2 the minimum NEP will therefore be $5.5 \times 10^{-11}\text{ W Hz}^{-1/2}$ if $T = 300$ K. If the detector and its surroundings are cooled to 1.5 K, the minimum NEP falls to $9.8 \times 10^{-17}\text{ W Hz}^{-1/2}$.

This minimum possible NEP does not take into account the absorption of any signal power by the detector. When the signal power is significantly higher than the background level radiated by the detector's immediate surroundings, the photon shot noise level will also rise. This means that the signal-to-noise ratio observed at a particular signal power level does not simply improve in proportion with any increase in the signal power.

Noise arises in the detector element, in the biasing electronics and in the signal-amplifying system. The actual NEP of a practical system will therefore always be higher than the minimum level set by photon–phonon noise. The total noise level in a practical bolometer system may be calculated by adding together the individual noise powers generated in each part of the system.

Reducing the temperature of a bolometer and its surroundings both improves its responsivity and lowers the NEP. The level of thermal electronic noise produced within the bolometer element and its biasing components may also be reduced by cooling. In some cases it is also possible to cool the amplifiers used with the bolometer, hence reducing any noise that they may contribute.

6.2 Types of Bolometer

Here we shall consider three examples of bolometric detectors which have been widely used to detect millimetre-wave signals; the *Golay detector*, the *bismuth bolometer* and the *indium antimonide* (*InSb*) hot-electron bolometer.

The Golay detector and bismuth bolometer are normally used at room temperature as convenient sensors with moderate sensitivity. Hence they do not require a cryostat system. Figure 6.4 illustrates how these two sensors work.

The Golay cell consists of an absorbing film placed inside a pressure cell. Signal power absorbed by the film is transferred by conduction to the gas inside the cell. Hence the gas pressure inside the cell will alter in response to a change in the input signal power level. In most Golay detectors, changes in pressure are detected by observing the flexing of a thin cell wall. This is done by silvering the thin part of the wall and measuring the deflection of a reflected beam of light. An alternative to the Golay detector is a similar

Bolometric Signal Detection

form of cell which uses a conventional microphone to measure the change in gas pressure.

The main disadvantages of the Golay cell are its poor response time and its sensitivity to mechanical vibration (microphony). A typical Golay detector requires the order of a second to respond fully to a change in the input signal power level. In order to prevent the fragile cell from being destroyed by changes in atmospheric pressure or ambient temperature, a small vent is generally included to allow the cell pressure to come to equilibrium with the external pressure. Hence their response is transient.

Golay detectors have been commercially available for many years and, in some quarters, their relative age is associated with obsolescence. They are, however, still useful when very high sensitivity is not required. A typical Golay detector will have a responsivity of 10^3-10^6 V W^{-1} and an NEP of around 10^{-10} W Hz$^{-1/2}$. They can therefore be used to measure power fluctuations down to a few tens of nanowatts at modulation frequencies up to 10–20 Hz.

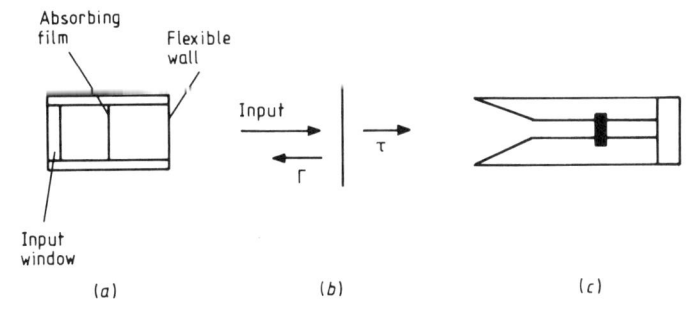

Figure 6.4 Three examples of room-temperature bolometric detector arrangements. (a) Golay pressure cell. (b) Bismuth resistive film. (c) Small bismuth bolometer mounted on E-plane line and antenna system.

Golay detectors often respond to a very wide range of signal frequencies, from the visible to the millimetre-wave region. This is useful in wide-band measurements but can cause problems such as unwanted sensitivity to room lighting. Filters can help to prevent these problems. The cell wall of a sensitive Golay detector is very fragile. A sharp impact—or even a sudden change in the signal power level—may rupture the cell wall. These problems can be avoided by handling the detector with care, filtering the input and avoiding any sudden change in the input level.

The input window on a typical Golay cell is quite small, perhaps a few millimetres across. This is because they are generally intended for use in the near infrared where this size is convenient. In order to couple millimetre-wave power onto the cell efficiently, an appropriate feed horn is required.

Two general types of bismuth bolometer are used to detect millimetre-wave radiation, both illustrated in figure 6.4.

Figure 6.4(b) shows a thin film of bismuth evaporated onto a dielectric sheet. When the effect of the dielectric (which is present simply as a physical support) is neglected, the bismuth film behaves as a resistive sheet onto which we may direct a millimetre-wave free-space beam. A pair of electrical contacts at opposite edges of the film can be used to apply a small current and to measure the film resistance.

The electrical resistivity of bismuth is quite high (10.7 $\mu\Omega$ m at 0 °C) compared with most normal metals (e.g. copper has an electrical resistivity of 0.15 $\mu\Omega$ m at 0 °C) and the temperature coefficient of bismuth (4.2 mΩ m °C^{-1}) is moderately high. A thin film of bismuth can therefore be used to absorb signal power and the change in temperature sensed by measuring the change in resistivity.

The behaviour of a thin metallic film placed perpendicular to a signal beam can be analysed using the transmission line ideas outlined in chapter 5. The incident beam behaves as a signal propagating along a line whose characteristic impedance Z_c is equal to the impedance of free space. The metal film can be regarded as a shunt load whose resistance is equal to the resistance 'per square' of the metal film. The units of area do not in fact matter when the resistance per unit area of a film is calculated or measured, i.e. the amount of resistance in ohms per square centimetre or per square inch (or even per acre!) of a uniform film is always the same value.

The signal reflectivity Γ and transmissivity τ of a film whose resistance per square is Z will be

$$\Gamma = \frac{Z' - Z_c}{Z' + Z_c} \tag{6.29}$$

where

$$Z' = \frac{ZZ_c}{Z + Z_c} \tag{6.30}$$

and

$$\tau = 1 + \Gamma. \tag{6.31}$$

The fraction A of the incident signal power absorbed by the film will hence be

$$A = 1 - |\Gamma|^2 - |\tau|^2. \tag{6.32}$$

For a resistive film, Γ and τ are both real; hence we may write that

$$A = -2(\Gamma + \Gamma^2). \tag{6.33}$$

At first glance, equation (6.33) appears to imply a negative power absorption. However, from equation (6.30), it can be seen that $Z' \leqslant Z_c$. It follows from this that the reflectivity must be in the range $-1 \leqslant \Gamma \leqslant 0$, i.e. Γ is negative and the power absorptivity A will be positive or zero. The maximum value of A occurs at $Z = \tfrac{1}{2} Z_c$, which corresponds to a resistance per square of around 180 Ω. This produces a peak power absorptivity of 0.5.

Bismuth films of this type are moderately simple to make by vacuum evaporation onto thin dielectric sheets. Although they are fragile and must be protected from aging effects caused by exposure to air, they work well over a very wide signal frequency range. They also can be used to absorb signal power almost irrespective of the beam pattern directed onto the film. Hence they are useful for measuring the signal power of a beam whose pattern is not well defined.

The amount of power absorbed may be increased by providing a suitable impedance-matching or reflection- and transmission-cancelling arrangement around the absorbing film. For example, a mirror may be placed behind a film and used as an optical back-short. The film and mirror then appear to the incoming signal as a terminating load instead of a shunt across a continuing transmission line. If the film resistance per square is now chosen to be equal to Z_c and the mirror is a quarter wavelength from the film, the termination is matched to free space and all the incident power will be absorbed. Although the improved absorptivity is useful, the addition of a mirror back-short 'tunes' the detector and the signal absorptivity now has a frequency variability which depends upon the mirror–film spacing.

A large bismuth film detector will be somewhat less sensitive than a Golay cell if it is designed to have a similar response time. It will not be microphonic but may need to be protected from temperature fluctuations in the surrounding air. A typical film, designed to have a response time of just under a second, will be useful for measuring power levels around a few milliwatts or more.

In figure 6.4(c) an alternative method for using bismuth bolometers in which an E-plane antenna is used to couple a free-space beam into an E-plane transmission line is illustrated. A piece of bismuth (which is small compared with the signal wavelength with power coupled into it via some form of antenna and mounting circuit) is placed across the line. Hence some of the signal power transmitted along the line will set up a current in the bismuth and the dissipated power may be detected. A very small bismuth element may be treated as a shunt placed across the transmission line. A larger element may need to be considered as a length of 'lossy' transmission line. In either case the system may be designed using normal transmission line arguments. In the illustrated example a back-short is placed behind the bismuth element to increase the power absorptivity at a particular signal frequency.

This arrangement differs from the simple large film in that it now possesses a definite antenna pattern. The responsivity may be considerably greater than that of a large film. This is because the signal power is coupled into a much smaller thermal capacity and tends to produce a greater temperature change. This in turn means that a higher thermal conductivity may be used to carry heat away from the detector, reducing the response time without producing an unacceptable degradation of the responsivity.

The two forms of bismuth bolometer illustrate two differing approaches to device construction. The large film acts as a 'distributed' or 'bulk' system where power is beamed directly into the active material. The small element acts as a 'two-terminal' device into which power is coupled as a current and potential between two end contacts. One approach is typical of optical systems, and the other of electronic systems. The main distinction between these approaches is whether the device is either large or small compared with the signal wavelength. In the millimetre-wave region the radiation wavelength and device size are often comparable. Either approach may then be used and the underlying physics may be seen to be the same.

Neither the Golay detector nor the bismuth bolometers can be cooled to very low temperatures. The gas in a Golay cell would condense if cooled to liquid-helium temperatures and the resistance properties of bismuth becomes unsatisfactory. The InSb bolometer, however, *must* be cooled in order to be used as a far-infrared detector. This is because its electronic properties only become satisfactory at low temperatures.

In many types of bolometer, incident (millimetre-wave) energy is primarily absorbed by electrons which are closely bound to particular lattice atoms. The absorbed energy and momentum are therefore quickly transformed into lattice vibrations (phonons), i.e. the crystal lattice warms up.

However, in a cooled InSb sample, as many of the electrons are weakly coupled to the crystal lattice (generally referred to as an *electron gas*), they are able to move fairly easily around the crystal. Thus, any energy absorbed by the InSb will tend to change its electrical resistance; this allows us to use the crystal as a bolometer.

Also the input signal power raises the mean energy of these electrons just as if the crystal lattice temperature had been increased. The electrons now act as if the crystal were hotter than its actual physical temperature. This is referred to as a *hot-electron* effect. The crystal lattice acts as the heat sink because the weak coupling between electrons and lattice assumes the role of a small thermal conductance and the thermal capacity of the electron gas is also very small.

The combination of low thermal capacity and low conductance produces a bolometer which has a high responsivity and a short response time. A typical InSb bolometer, cooled to around 3 K, will have an intrinsic response time of the order of a microsecond, a responsivity of perhaps 10 kV W^{-1} and an NEP well below 10^{-12} W Hz$^{-1/2}$. The sensitivity and noise performance may be improved by cooling below 1 K. InSb detectors

Bolometric Signal Detection

have also sometimes been used at higher temperatures. This degrades the responsivity and noise performance but reduces still further the response time. At temperatures above 20 K the response time falls below 0.1 μs.

6.3 Phase-sensitive Detection and Background Subtraction

In many cases we need to measure a very small signal power level in the presence of noise. We may also wish to measure a signal which is dominated by an unwanted level of 'background' power coming from other sources in which we have no interest. As an extreme example of these problems we can consider the problems which arise when an astronomer wishes to use a bolometer to measure the millimetre-wave signals radiated by a cloud of gas and dust some considerable distance from the earth.

Signal power can be collected and directed onto a bolometer using an antenna system. The input signal power level collected in this way is usually very small. An antenna system on the earth's surface must look through our atmosphere. This will absorb some of the signal before it can reach the detector. The atmosphere in the line of sight may also radiate far more power onto the detector than is received from the astronomical source. To make matters worse, the atmospheric absorption and emission fluctuate randomly from moment to moment.

Essentially all forms of active device, detectors, amplifiers, etc, suffer from *excess* noise. The forms of random noise discussed earlier all have 'white' frequency spectra, i.e. the amount of noise power per hertz bandwidth does not depend upon the frequency. Unfortunately, one of the most common forms of excess noise is $1/f$ *noise*. This, as its name implies, has a power spectrum which varies inversely with frequency. Atmospheric fluctuations also exhibit a $1/f$ spectrum at low frequencies.

Some detectors (e.g. the Golay cell) are deliberately designed in a way which makes them unsuitable for measuring fixed signal levels. Even when this is not the case it is desirable to find a way of measuring fixed or slowly changing signals which overcomes—as far as possible—these difficulties.

There are two common ways of dealing with these problems.

(i) We can *modulate* the signal, changing a steady quantity into an alternating one.

(ii) We can *subtract* any background effects from the signal before measurement.

Phase-sensitive detection (PSD) systems employ both modulation and subtraction techniques and are often used to measure small signals in the presence of noise and unwanted background effects.

Figure 6.5 illustrates a typical phase-sensitive detection system being used to measure the signal power from a weak source. In this example a *beam chopper* is used periodically to prevent signal power from reaching the detector. A common form of chopper is a toothed wheel which is rotated so that its blades pass through the source–detector beam. The system shown in figure 6.5 has a *reflecting* chopper, i.e. when a blade is placed in the beam the detector sees power from a *comparison* source. As the chopper rotates, the signal power level reaching the detector switches rhythmically between the levels produced by the source of interest and the comparison. In the arrangement illustrated, another mirror is used to ensure that the comparison source is seen against a similar background to that of the signal source.

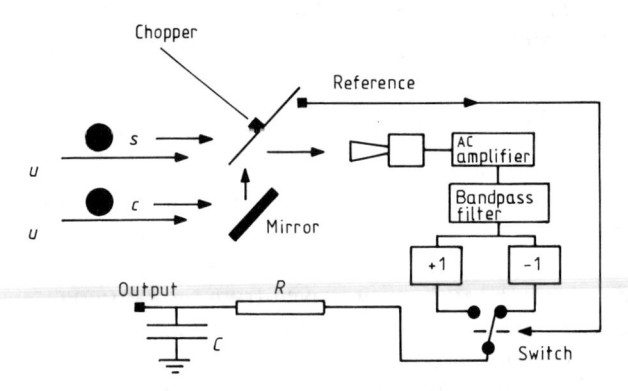

Figure 6.5 Phase-sensitive detection (PSD) system.

In some cases the chopper will be made of a material which strongly absorbs (and hence emits) millimetre-wave radiation. The chopper blades themselves then form the comparison source and radiate an amount of power determined by their temperature. In some other cases a reflecting chopper is used without a specific comparison source. Under these circumstances the detector will receive some power from the surroundings as seen by reflection from the chopper blades. Sometimes it is possible to dispense altogether with a physical chopper and simply to switch the source on and off.

Whatever the details of the arrangement, the result is to produce a power level at the detector which switches in a predetermined way between the signal level that we wish to measure and a specific comparison level. If the source power level is s, the comparison power level c and the common background power level is u, the magnitude A of the periodic change in detector output will be

$$A = (s + u) - (c + u) = s - c. \quad (6.34)$$

By chopping or switching the signal before the detector we have produced an AC output whose size depends upon the signal s but *not* upon any general background level u. This result remains true even when $s \ll u$, provided only that the background level is not so high as to overload the detector. We may now measure the size of the AC signal and use this as a measure of the signal power level.

The nominally steady input level has been *modulated* at a particular chopping frequency f_c. The detector output can now be enlarged using AC amplifiers and passed through a bandpass filter which lets through only a limited range of frequencies centred on f_c. Any noise fluctuations produced by the background, detector or the amplifiers at frequencies significantly different from f_c will be rejected by the filter.

Only noise in the frequency range passed by the filter will affect the output measurement. By reducing the bandpass range we can lower the amount of noise which passes through the filter. This also means that, if we choose a chopping frequency which is above the region where $1/f$ noise is a problem, we can stop any of this extra noise from appearing in the final measurement.

Unfortunately, as the filter bandwidth is reduced, we begin to encounter some other problems. The most fundamental of these arises because of the fundamental properties of a bandpass filter. A filter designed to pass only a narrow frequency range behaves essentially as a *resonant* filter. We can define the *bandwidth B* of a filter as

$$B = f(\text{max}) - f(\text{min}) \qquad (6.35)$$

where $f(\text{max})$ and $f(\text{min})$ are the maximum and minimum modulation frequencies which will pass through the filter. The time taken for a resonance to build up or decay will depend upon $1/B$. Hence, the narrower that we make the filter bandwidth in order to reduce noise, the longer the filter output takes to respond to a change in the AC level at its input.

For various practical reasons, filters where $B \ll f_c$ are difficult to make. Also we may find that, if B is too small, any small changes in the chopping frequency may cause f_c to move outside the range $f(\text{min}) < f_c < f(\text{max})$, and the signal that we require may not always pass through the filter. Phase-sensitive detection provides a method for dealing with these problems.

In the system shown in figure 6.5 the amplified AC signal is passed to a parallel pair of amplifiers whose gains are $+1$ and -1. A *reference* signal is taken from the beam chopper and used to operate a switch which selects which amplifier to connect to the output. The signal is then passed from the switch to a capacitor of capacitance C via a resistor of resistance R. The combination of R and C acts as a time constant which averages the selected signal voltages over the time $\tau = RC$. This *smooths* the switched signal to produce an output level which takes a time τ to change significantly.

Provided that the switch is operating correctly *in phase* with the chopping between source and comparison, then the AC signal is rectified by the switching process. If the chosen time constant τ is much greater than $1/f_c$, then the voltage on the capacitor will be a steady level whose size is proportional to $s - c$. The overall system has taken a steady input, converting it to an AC signal for amplifying and processing and then changed it back into a steady signal for final measurement.

The reference frequency must be equal to the chopper frequency f_c as both are generated by the same device. It is, however, possible for there to be a phase difference between the two. The output voltage on the capacitor will depend upon the phase between reference and signal. If the relative phase of the reference is changed by $180°$, the switch setting will be reversed at any particular part of the signal cycle. This produces an output which is inverted; hence the voltage on the capacitor has the same magnitude as before but is opposite in sign. If the reference and signal are $90°$ out of phase, then the inverted and non-inverted parts of the signal are equal and opposite, i.e. the averaged voltage is *zero*.

In general, the smoothed output voltage will be proportional to

$$(s - c) \cos \phi \tag{6.36}$$

where ϕ is the phase difference between the reference and the AC signal.

Any noise at a frequency f_n which is close to the frequency f_c may be regarded as being proportional to

$$\cos(2\pi f_c t + \phi) \tag{6.37}$$

where

$$\phi = 2\pi(f_n - f_c)t \tag{6.38}$$

i.e. noise at a frequency near the chopping frequency can be regarded as being at the chopping frequency but having a phase relative to the reference which varies linearly with time. This noise produces a switched output level which varies as

$$\cos[2\pi(f_n - f_c)t]. \tag{6.39}$$

The varying output is then averaged by the output *RC* circuit. This means that, when $2\pi(f_n - f_c) > 1/\tau$, the noise has little or no effect upon the output voltage which appears on the capacitor.

The phase-sensitive detector acts as a filter which suppresses the effects of noise outside the range $f_c \pm 1/2\pi\tau$. The arrangement has three advantages over simply using a bandpass filter. Firstly, the arrangement continues to work even if f_c changes; secondly, the *RC* filter is simple and time constants up to the order of 100 s (equivalent to a filter bandwidth of less than 0.01 Hz) can easily be produced; finally, the filter is *phase* sensitive.

The phase-sensitive nature of the arrangement means that even noise *at* the chopping frequency f_c will be prevented from having an effect upon the output voltage if the noise is 90° out of phase with the reference (and the wanted signal).

There will always be some noise at the chopping frequency. Being random, its phase will vary unpredictably. We can treat this noise as if it were composed of *two* random components which, on average, have similar power levels, one in phase with the reference and one in quadrature. The quadrature component cannot produce any output; hence the phase-sensitive detector essentially rejects half the noise power at the chopping frequency f_c.

Phase-sensitive detection systems are simple to build, are flexible and provide an excellent way of dealing with background and noise problems. They are therefore widely used. A range of slightly different practical techniques have arisen in various fields and have been given various names—an example being *sky chopping* or *beam nodding* used by astronomers to deal with the situation mentioned at the start of this section. Here the telescope or antenna is moved so that the source is switched in and out of the detector beam. Provided that they are not too far apart, the atmospheric emission in the 'source' and 'off-source' directions is much the same. A phase-sensitive detection technique referenced to the antenna beam direction can then be used to observe the source power and to suppress the effects of atmospheric absorption. A similar technique is also used by radio astronomers who refer to the method as *Dicke switching*.

Phase-sensitive detectors act as *comparators*, i.e. the output depends upon the difference in signal level s produced by the source and a comparison power level c. This means that the signal measurement can only be as accurate as our knowledge of the level that we expect from the comparison. (Note that the comparison is sometimes called a *reference source*. This usage has been avoided here as it is easily confused with the reference taken from the chopper to control the output switch.)

Most practical measurements are, in fact, comparisons rather than 'absolute' measurements. We measure the length of some object by comparing it with a ruler, and a time by counting the ticks of a clock. When buying a gallon of petrol, we are less concerned with the absolute definition of a gallon than with feeling certain that this gallon is the same as that at another filling station a few days ago. Similarly, our comparison power level c should be a known standard which can be reproduced or defined with whatever accuracy we require.

In some circumstances the necessity for a reliable comparison may prove a problem. In other cases the comparison forms the basis of an important measurement technique, that of *null measurement*.

In general, to measure s we must know the detector responsivity and the gain of the signal-processing system. Any errors in our knowledge of these factors will produce an error in our final measurement. If, however, we can adjust c in a controlled manner and set $c = s$, then the output will be zero irrespective of the detector responsivity and system gain. A high responsivity and gain will mean that any small difference between s and c will register at the output of the system, but the accuracy of our measurement does not depend upon a prior knowledge of these factors. The reliability of the result depends only upon the accuracy of the *null* produced between s and c, and our knowledge of the comparison power level c.

Chapter 7

Mixers and Heterodyne Detection

7.1 Signal Mixing in Diodes

Bolometric signal detectors respond simply to the power level of a signal. *Heterodyne* detection systems produce output which contains information about the frequency distribution of incident signal power. They are therefore of particular value when we wish to determine the power frequency spectrum of a signal. A complete heterodyne detection system is usually called a *heterodyne receiver*. The detecting element in a heterodyne system is called a *mixer*. This is because it 'mixes' together two input signal frequencies to produce an output at another frequency. One of the simplest, and most commonly used, forms of mixer is a diode.

If we alter the voltage applied to a resistor, we find that the resulting current varies in proportion to the voltage. The resistor obeys *Ohm's law* and is described as an *Ohmic* device. The voltage V and current I are related by the simple expression

$$V = IR \tag{7.1}$$

where R is the resistance of the chosen resistor. A graph of V against I for such a resistor will be a straight line of slope R.

An electronic diode does not obey Ohm's law. If we plot V against I for a diode, we discover that the resulting line is curved. A diode exhibits a *non-linear I–V curve*. The behaviour of the device as a mixer depends upon this non-linear electronic property.

Most common diodes exhibit an I–V curve where the device current depends upon the *sign* of the applied voltage as well as its magnitude. A voltage applied in one direction generates almost no current unless the voltage is so high that the diode 'breaks down'. A voltage applied in the other direction produces a current which increases roughly exponentially with increasing applied voltage, i.e. the I–V behaviour of a typical diode may be represented as

$$I \simeq 0 \qquad V < 0 \qquad (\textit{reverse direction}) \qquad (7.2)$$
$$I \propto \exp(\beta V) - 1 \qquad V \geq 0 \qquad (\textit{forward direction}) \qquad (7.3)$$

where we have adopted the general convention that the main current flow occurs when a 'positive' voltage is applied and that is regarded as being the 'forward' direction of current flow.

Here we are interested in the behaviour of the diode when used in the forward direction. Equation (7.3) may be rewritten as a power series

$$I = a_1 V + a_2 V^2 + a_3 V^3 + \cdots \qquad (7.4)$$

with appropriate values of the coefficients a_n. For simplicity, here we shall assume that only the V^2 coefficient is significant, i.e. we shall assume that the diode I–V curve is of the form

$$I = aV^2 \qquad (7.5)$$

(where a is equivalent to the a_2 coefficient in equation (7.4)). This considers the diode to behave as what is called a *square-law device*.

The assumption of a square law-device turns out to be fairly good under most circumstances. If necessary, the same arguments as can now be illustrated using the square-law assumption can be repeated using a more complex I–V expression. The mathematical expressions become more involved when this is done, but the final result is similar in most practical cases.

Consider now the system illustrated in figure 7.1. Figure 7.1(a) represents a pair of bandpass filters being used to couple two signals, at the frequencies f_s and f_1, into a diode. Filter 1 is designed to behave as a short circuit at the frequency f_s and as an open circuit at the frequency f_1. Filter 2 is designed to be a short circuit at the frequency f_1 and an open circuit at the frequency f_s. A steady *bias voltage* V_0 is also applied to the diode via the inductance L.

If the two input signals, produced by the generators attached to the filters, are

$$V_1 = V_s \sin(2\pi f_s t) \qquad V_2 = V_1 \sin(2\pi f_1 t) \qquad (7.6)$$

then the total voltage seen at the diode will be

$$V = V_0 + V_s + V_1. \qquad (7.7)$$

The diode current will therefore be

$$I = a[V_0 + V_s \sin(2\pi f_s t) + V_1 \sin(2\pi f_1 t)]^2. \qquad (7.8)$$

Mixers and Heterodyne Detection

This expression may be rewritten in the form

$$I = aV_0^2 + 2aV_0V_s \sin(2\pi f_s t) + 2aV_0V_1 \sin(2\pi f_1 t)$$
$$+ \tfrac{1}{2}aV_s^2\{1 - \cos[2\pi(2f_s)t]\} + \tfrac{1}{2}aV_1^2\{1 - \cos[2\pi(2f_1)t]\}$$
$$+ aV_sV_1\{\cos[2\pi(f_s - f_1)t] - \cos[2\pi(f_s + f_1)t]\}. \qquad (7.9)$$

If we consider equation (7.9), it can be seen that the diode current has oscillatory components at the frequencies $2f_s$, $2f_1$, $f_s + f_1$ and $f_s - f_1$, as well as at the two input frequencies f_s and f_1. The diode thus behaves as a *frequency conversion* device, transferring some of the power input at the frequencies f_s and f_1 to other frequencies. In general, a device which has a non-linear $I-V$ curve which is more complex than the simple square-law case will convert power to a range of frequencies $mf_s \pm nf_1$, where m and n are integers.

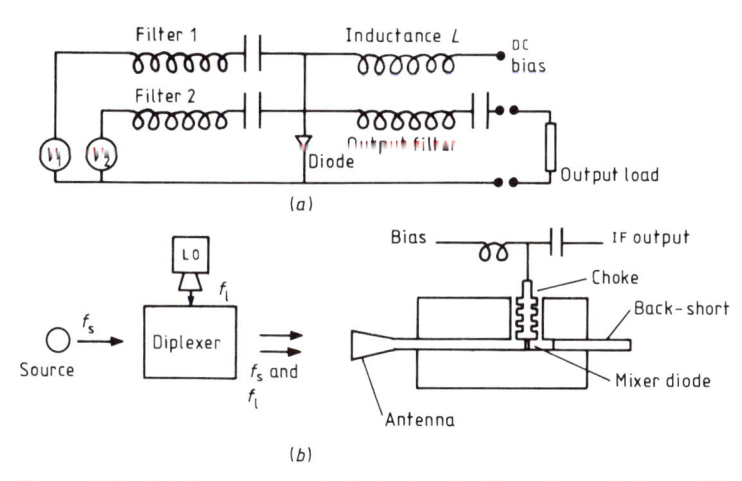

Figure 7.1 (a) Arrangement of electronic filters intended to efficiently couple signals into and out of a mixer diode. (b) Equivalent system using optical and waveguide methods.

None of these newly created frequencies may pass through the input filters or the bias inductor. We can, however, extract the power generated at these frequencies via the output filter. In most practical cases, we are interested mainly in the *difference frequency* $f_s - f_1$, and the output filter will be designed to couple this output so as to produce a signal

$$V(\text{out}) \propto V_s V_1 \cos[2\pi(f_s - f_1)t] \qquad (7.10)$$

across the output load (which will generally be the input impedance of an amplifier).

Figure 7.1(*b*) is a schematic diagram of a typical millimetre-wave mixer system. The mixer diode is mounted as a shunt across a waveguide transmission line and a back-short is placed behind the diode to improve power matching from the waveguide into the diode. The waveguide is coupled to an antenna which collects incident millimetre-wave power.

The input filters 1 and 2 in figure 7.1(*a*) are replaced by an optical *diplexer* system. This takes signal power at the frequency f_s from a source of interest and couples it efficiently into the mixer's antenna. It simultaneously takes the output from a *local oscillator* (LO) at the frequency f_l and couples this into the antenna. DC bias is passed into the diode via a *bias choke* which is designed to allow DC into the device and to permit the difference frequency to escape but prevents any power at the signal or LO frequencies from escaping. The bias supplied to the mixer is separated from the difference-frequency output by a suitable filter—often called a *bias tee* because many of these filters are 'T' shaped.

The input signal and LO frequencies are typically around 100 GHz or more; however, by selecting an LO frequency which is reasonably similar to that of the signal, the output difference frequency may be a few gigahertz or less. The system thus acts as *down-converter*. An input signal $V_s \sin(2\pi f_s t)$ will generate an output $a V_s V_l \cos(2\pi f_0 t)$, where $f_0 = f_s - f_l \ll f_s$. This output signal can now be processed using conventional electronic techniques even though the actual input signal frequency f_s is extremely high.

The output frequency f_0 is usually amplified, filtered, etc, before finally being measured. It is conventional to refer to this frequency as the intermediate frequency (IF) and to the output signal as the IF *signal* as it represents the signal information at a frequency intermediate between that of the initial input and the final measurement–analysis system which often converts it to a still lower frequency.

Provided that the LO frequency is known and well defined, then the signal frequency may be deduced by measuring the frequency f_0 of the IF output from the mixer. This is an important difference between heterodyne receivers and bolometric detectors which do not inherently supply any information regarding the signal frequency.

Most real signals do not, in fact, consist simply of a single component of fixed amplitude and frequency. More commonly the input signal will be composed of a set of frequency components or a continuous spectral distribution.

If the argument employed above is extended, it can be seen that an input signal of the form

$$S(t) = \int_{f(\min)}^{f(\max)} A(f) \exp[-j(2\pi ft + \phi(f))] \, df \tag{7.11}$$

which represents a continuous spectrum between the frequencies $f(\max)$ and $f(\min)$ will, when mixed with a LO frequency f_l produce an IF output of the form

$$S'(t) = \int_{f'(\text{min})}^{f'(\text{max})} A'(f') \exp[-j(2\pi f't + \phi(f'))] \, df' \quad (7.12)$$

where

$$A'(f') \propto A(f) \quad (7.13)$$
$$\phi(f') = \phi(f) \quad (7.14)$$

and

$$f' = f - f_1 \quad (7.15)$$

i.e. the output IF spectrum has an amplitude–frequency spectrum which is proportional to that of the input but is shifted down in frequency by f_1. It can also be seen that the relative phases of output IF components are the same as those in the initial signal. In principle, the output IF signal contains all the information which existed in the original input. The IF output from a mixer may therefore be amplified and passed to a suitable data analysis system which will be able to recover detailed spectral information about the input signal. This is a particularly useful capability of heterodyne detection systems.

In practice, IF output at a frequency f_0 may be produced by a signal at either of *two* frequencies

$$f_s = f_1 \pm f_0 \quad (7.16)$$

This is because $\cos\theta = \cos(-\theta)$. Hence the output generated according to equation (7.10) will have the same frequency for either of these two possible inputs. For a given LO frequency, a mixer which produces output over the IF range from $f(\text{min})$ to $f(\text{max})$ will therefore respond to input signals in the range from $f_1 + f(\text{min})$ to $f_1 + f(\text{max})$ and to signals in the range from $f_1 - f(\text{max})$ to $f_1 - f(\text{min})$. There are *two* bands of receivable signal frequencies, placed either side of the LO frequency.

The band from $f_1 + f(\text{min})$ to $f_1 + f(\text{max})$ is conventionally referred to as the *upper sideband*. The band from $f_1 - f(\text{max})$ to $f_1 - f(\text{min})$ is referred to as the *lower sideband*. A receiver system which responds to both is called a *double-sideband receiver*.

In some cases the existence of two distinct sidebands does not matter. The input signal may, for example, be uniform over a wide frequency range. The presence of two input bands then simply allows more signal power to be detected, making the detector more sensitive. In some other cases the source output will already be fairly well known and consists of components which fall into only one sideband. More generally, however, the chance that a given LO output may be produced by an input signal at either of two possible frequencies gives rise to ambiguities in a measurement.

The simplest way of dealing with this problem is to place a signal filter between the source and the mixer. This filter is arranged so as to pass only one sideband without loss and strongly rejects the other unwanted sideband. The use of such a filter produces a *single-sideband* receiver system. The required frequency range in such a system is sometimes called the *signal sideband*. The other unwanted sideband is sometimes called the *image*.

The ability of a single-sideband heterodyne receiver to measure input signal powers and frequencies accurately depends upon the frequency and amplitude stability of the LO employed. Any uncontrolled or random changes in f_l or V_l are multiplied into the output, reducing its value as a measure of the input signal. It is therefore important to provide a stable LO input. The LO generator will, itself, often generate noise at frequencies which fall into the signal and image bands. In order to obtain good performance this should be as low as possible. It is also useful, whenever possible, to employ a diplexer which prevents LO output in these bands from reaching the mixer.

Heterodyne systems can provide frequency measurements of very high accuracy and resolution. When these systems are used at frequencies up to a few hundred gigahertz, it is possible to measure frequencies with an accuracy and precision of a few kilohertz (i.e. 1 part in 10^8) or better if required.

Conventional mixer diodes are *junction* devices. Their properties arise as a consequence of the behaviour of charge carriers in the region surrounding the junction between two different materials. Most of the diodes used at radio frequencies are simple *p–n junctions* usually manufactured from doped silicon. At millimetre-wave frequencies, mixers are more commonly made using *Schottky barriers* which consist of a junction between a metal and a semiconductor (usually gallium arsenide (GaAs)).

Various other forms of mixer diodes have also been used. Examples include *semiconductor–insulator–semiconductor (SIS) diodes* and *Josephson junctions*, which depend upon the properties of semiconductors and very thin barriers. The active part of all of these devices is very small compared with millimetre wavelengths, a typical junction being of the order of a micron or just a few microns in diameter. Hence they must be used as 'two-terminal' devices. Power must be coupled into the active device via a pair of attached electronic contacts.

Problems arise in using junction devices because the thickness of the active junction region is very small—generally somewhat less than a micron. As a result the device capacitance per unit area is significant. Given the high frequencies entailed in millimetre-wave signals this means that the junction capacitance will act as a shunt, preventing the device from operating as required, unless the area is kept very small. A small device cross section, however, implies that Ohmic resistive losses may become a problem. Figure 7.2 represents an 'idealized' form of a junction diode. The arrangement in figure 7.2(*a*) represents a device of cross sectional area A and overall length

Mixers and Heterodyne Detection

h. The active *depletion region* is of width d. For simplicity it is assumed that the bulk of the materials on both sides of the active junction region has a resistivity ρ and a relative permittivity ε. The junction region is assumed to share the same permittivity but has a somewhat higher resistivity ρ'. Ohmic contacts, of negligible resistivity, are placed covering the ends of the device and are used to couple signals in and out of the device via external wires.

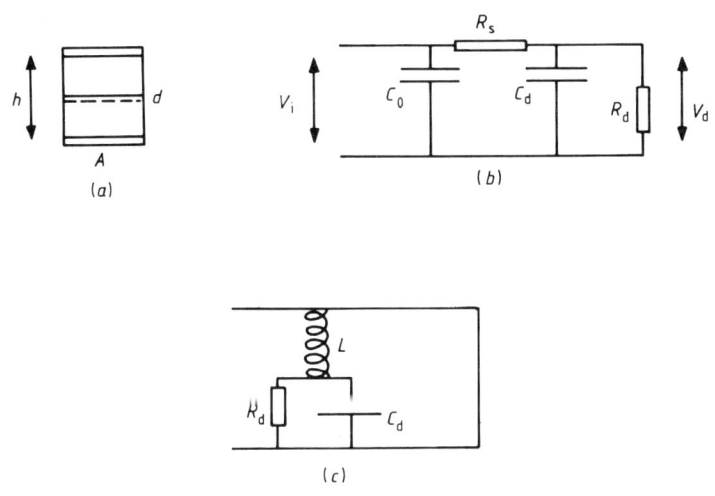

Figure 7.2 (a) Simple physical model of a junction diode. (b) Electronic model of a diode. (c) Electronic model of a diode and the surrounding circuit.

This physical arrangement may be represented as an electronic circuit of the form shown in figure 7.2(b), where

$$R_s = \frac{h\rho}{A} \qquad R_d = \frac{d\rho'}{A} \qquad (7.17)$$

$$C_0 = \frac{\varepsilon A}{h} \qquad C_d = \frac{\varepsilon A}{d}. \qquad (7.18)$$

In a real device the width d of the depletion region and its effective resistivity ρ' will depend upon the applied voltage. The resistance R_d and capacitance C_d do not therefore have unique values. Here we shall assume that R_d corresponds to the *gradient* $\delta V/\delta I$ of the diode I–V curve. The capacitance C_d is assumed to be $\delta q/\delta V$, where q is the charge stored in the device capacitance. Both of these quantities will depend upon the chosen bias voltage and act as the effective resistance and capacitance for small voltage changes about the chosen level.

This simple model should not be taken as an accurate representation of any particular form of junction device. The physical details of a real device differ in various respects and its behaviour is considerably more complex. However, the model is useful in demonstrating some of the basic limitations of junction devices.

The non-linear resistance of the device resides in the variable nature of R_d. Hence an input signal V_i applied between the end contacts must be coupled without significant loss to R_d if the device is to work efficiently. Between the input terminals and the 'active' diode resistance R_d the combination of R_s and C_d form a time constant which acts as a low-pass filter.

The resistance R_s is generally called the device *spreading resistance* as it represents the resistance through which any current must be distributed or spread to reach the active region. Provided that the device is well designed and $R_s < R_d$ we can say that the signal voltage V_d across R_d will be such that

$$|V_d| = \frac{|V_i|}{\sqrt{1 + (2\pi f \tau)^2}} \qquad (7.19)$$

where f is the signal frequency and

$$\tau = R_s C_d = \frac{\varepsilon \rho h}{d} \qquad (7.20)$$

where we have assumed that h is somewhat larger than d. Hence, in order to be able to function efficiently, we require that

$$f < \frac{1}{2\pi \varepsilon \rho}. \qquad (7.21)$$

A more precise analysis of a particular device yields a result which, while differing in detail, behaves in much the same way. There is an upper frequency limitation imposed by the physical properties of the device materials which cannot be overcome by alterations in the size and shape of the device.

For GaAs materials used in millimetre-wave mixers, this problem means that devices cannot function well above approximately 1000 GHz. Practical difficulties in manufacturing real devices generally mean that it is difficult to approach this upper limit. Alternative choices of device materials and techniques can improve the situation, but all junction devices must face this basic problem.

A typical GaAs Schottky mixer with a diameter of the order of a micron will have a resistance of a few ohms and a capacitance of a few femtofarads (femto = 10^{-15}) under typical bias conditions. Whilst this is a small capacitance, it corresponds to an impedance of a few hundred ohms at millimetre-wave frequencies. A typical mixer diode therefore has a signal impedance typified by a low resistance shunted by a moderate capacitance. At higher frequencies the effects of capacitance begin to dominate the impedance of the device.

Mixers and Heterodyne Detection

A typical mixer will be mounted in a waveguide and a thin wire, or post, is placed across the guide so as to couple power into the device. This wire behaves essentially like an inductor placed across the guide in series with the mixer. Figure 7.2(c) represents the combination of such a wire and mixer diode placed as a shunt load across a transmission line. A back-short is used to improve signal matching into the mixer.

The characteristic impedance of a typical millimetre-wave waveguide is around 100–300 Ω. A typical mixer will have a resistance of the order of just a few ohms. Hence, by itself, the mixer is poorly matched to the transmission line. The situation may, however, be improved by careful choice of the inductance L and back-short position.

The properties of resonant circuits and their use to match impedances is considered in more detail in chapter 11. A similar approach can be employed here to establish the efficiency of signal coupling between the waveguide and a mixer diode. The combination of the diode, inductance and back-short will produce a total terminating impedance Z which can be defined from the expression

$$\frac{1}{Z} = \frac{1}{j\omega L + Z_d} + \frac{1}{jX_s} \quad (7.22)$$

where Z_d is the diode complex impedance and jX_s represents the back-short reactance.

We can also say that

$$\frac{1}{Z_d} = \frac{1}{\text{Re}(Z_d) + j\,\text{Im}(Z_d)} \quad (7.23)$$

where $\text{Re}(Z_d)$ and $\text{Im}(Z_d)$ are the real and imaginary parts of Z_d.

In order to match power ideally into the diode from a transmission line whose characteristic impedance is Z_c, we would require Z to be such that $\text{Re}(Z) = Z_c$ and $\text{Im}(Z) = 0$ at the frequency of interest. (Here $\text{Im}(Z)$ and $\text{Re}(Z)$ are used to represent the imaginary and real parts of Z.) If we combine and rearrange equations (7.22) and (7.23), these requirements can be seen to be equivalent to the conditions

$$(\text{Re}(Z_d))^2 + (\omega L + \text{Im}(Z_d))(\omega L + \text{Im}(Z_d) + X_s) = 0 \quad (7.24)$$

and

$$\text{Re}(Z_d)\, X_s^2 = Z_c [(\text{Re}(Z_d))^2 + (\omega L + X_s + \text{Im}(Z_d))^2]. \quad (7.25)$$

In practice it is difficult to satisfy both of these conditions and hence the diode coupling efficiency may be somewhat less than unity. The mixer then reflects some portion of the incident power. Furthermore, it can be seen that Z is strongly frequency dependent; hence it is generally only possible to produce a good impedance match over a restricted frequency range.

7.2 Bulk Mixers

It is very difficult to manufacture junction devices less than a micron or so in diameter. Any increase in device size will lower the resistance R_d and increase the capacitance C_d, making it harder to obtain good signal coupling. The problems of low resistance and shunt capacitance tend to rise with increasing signal frequency.

To overcome these problems it is necessary to replace the active junction with a mixer device which operates as a *bulk mixer*. To achieve this we require a material whose inherent resistivity depends upon the signal power level. To illustrate this we may consider the situation shown in figure 7.3. Here we have a piece of material which generates (or passes in response to an applied bias voltage) a current which varies proportionally to the input signal power level.

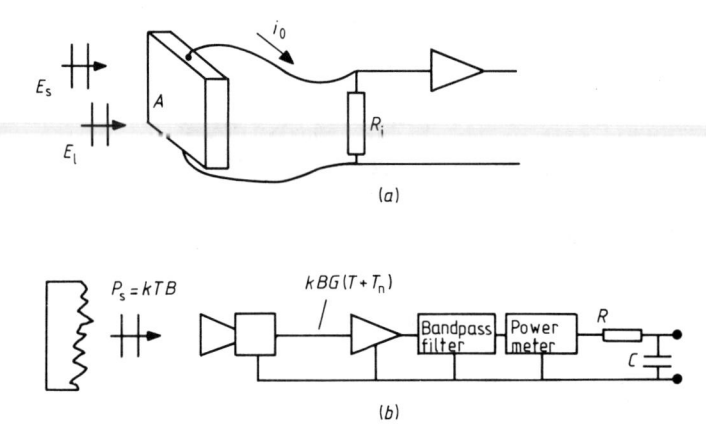

Figure 7.3 (a) Mixer crystal receiving two waves, E_l and E_s, and providing an output current, i_0. (b) Heterodyne receiver system.

In fact, it can be seen that most bolometric detectors which provide electrical output act in this way, i.e. they produce an output current which varies in proportion to the input signal power level. For this reason, most bolometric detectors can—in principle—be used as heterodyne mixers. In practice their use as mixers is normally severely limited for reasons which will become clear later.

Figure 7.3(*a*) shows a piece of material of surface area A illuminated with a signal field $E_s \sin(2\pi f_s t)$, and an LO field $E_l \sin(2\pi f_l t)$. Any current produced by the material is passed through the connecting wires attached to its ends.

The total power level $P(t)$ falling upon the material at any time t will therefore be

$$P(t) = \frac{[E_s \sin(2\pi f_s t) + E_1 \sin(2\pi f_1 t)]^2 A}{Z_0} \quad (7.26)$$

where Z_0 is the impedance of free space.

The material is assumed to behave as a *photoconductor*. Neglecting any current which flows when the input power is zero we may write that the input power generates a current

$$i(t) = \frac{e\eta P(t)}{h\nu}. \quad (7.27)$$

For simplicity it has been assumed that each incoming photon has an energy $h\nu$ equivalent to a frequency midway between the signal and LO frequencies (i.e. we assume that $\nu = \frac{1}{2}(f_s + f_1)$). In practice these two frequencies are generally very similar and this approximation is good.

The incoming power level then corresponds to $P(t)/h\nu$ photons per unit time. The factor η is an effective *quantum efficiency* which determines the probability that an arriving photon will produce a charge carrier (of charge e).

Combining the previous two expressions we can write that

$$i(t) = i_s + i_1 + 2\sqrt{i_s i_1} \cos[2\pi(f_s - f_1)t] \quad (7.28)$$

where

$$i_s = \frac{e\eta A E_s^2}{2 Z_0 h\nu} \quad (7.29)$$

and

$$i_1 = \frac{e\eta A E_1^2}{2 Z_0 h\nu} \quad (7.30)$$

plus some extra terms which correspond to current oscillations at the frequencies f_s, f_1, $f_s + f_1$, $2f_s$ and $2f_1$. When the input signal and LO frequencies are in the millimetre-wave region, these very-high-frequency components will not normally be passed along the wires to any connected measuring system. Here we shall assume that they do not influence any measurement of the current generated in the wire leads.

The steady currents i_s and i_1 are the values which would be obtained if just the signal or LO power were directed onto the detector element.

As in the case of a diode mixer, the combination of the signal and LO powers produces an oscillatory current at the difference frequency. The magnitude of this output current i_0 is

$$i_0 = \frac{e\eta A}{Z_0 h\nu} E_s E_1. \quad (7.31)$$

When the input signal and LO fields are in phase, the mean incident power will be proportional to $(E_s + E_l)^2$. When the fields are in opposition (180° out of phase), then the mean incident power will be proportional to $(E_s - E_l)^2$. The two input fields change their relative phase with the frequency $|(f_s - f_l)|$.

A given detector will take a specific time τ to respond to a change in the input power level. For most bolometers, this response time is of the order of a second or so. The output from the detector therefore indicates the mean incident power level, averaged over the time τ. The detector output will hence only oscillate significantly at the difference frequency when

$$|(f_s - f_l)| < \frac{1}{2\pi\tau}. \tag{7.32}$$

For a time constant of 1 s this requires the signal and LO frequencies to differ by less than 0.2 Hz. Hence, for almost all bolometric signal detectors, we find that the thermal response time means that the usable frequency bandwidth is very small, making the detector unsuitable for heterodyne reception under most circumstances.

An exception to this general finding is the InSb hot-electron bolometer. This has a response time of around a microsecond or less. As a result, InSb bulk mixers are used in some heterodyne receivers at very high frequencies. The response bandwidth of these mixers is typically around 1 MHz when cooled to a few kelvins above absolute zero.

7.3 Signal-to-noise Properties of Heterodyne Receivers

Figure 7.3(a) illustrates a mixer coupled to an amplifier whose input impedance is R_i. If the IF current produced by the mixer is $i_0 \cos[2\pi(f_s - f_l)t]$, then the mean power supplied to the amplifier will be

$$P_i = \langle i_0 \rangle^2 R_i \tag{7.33}$$

where $\langle i_0 \rangle^2$ is the mean squared current which, from the expressions given above, may be written as

$$\langle i_0 \rangle^2 = \frac{e^2 \eta^2 A^2}{2Z_0^2 h^2 \nu^2} E_s^2 E_l^2. \tag{7.34}$$

The mean signal power P_s and LO power P_l incident upon the mixer will be

$$P_s = \frac{E_s^2 A}{2Z_0} \qquad P_l = \frac{E_l^2 A}{2Z_0}. \tag{7.35}$$

Combining these expressions we obtain

$$P_i = GP_s \tag{7.36}$$

where

$$G = \frac{2e^2\eta^2 R_i}{h^2\nu^2} P_l. \qquad (7.37)$$

From equation (7.36) it can be seen that the IF power P_i supplied to the amplifier by the mixer is simply proportional to the signal power P_s received. The factor G which determines the ratio of these two power levels is conventionally called the *conversion gain* of the mixer.

The conversion gain depends upon both the load resistance R_i connected to the mixer and the incident LO power level P_l. In principle, we can assemble a mixer system which responds to an input signal by producing an output level $P_i > P_s$—such a system would amplify the signal as well as changing its frequency.

In practice, a range of problems often prevents us from producing a system with actual gain. A typical millimetre-wave mixer will have a conversion gain of less than unity, i.e. the gain is, in fact, a *loss*. A mixer is, by its very nature, a non-linear device. If the LO power level is increased sufficiently, the mixer will *saturate*. The output of the mixer then ceases to be power dependent and it no longer functions as a mixer. Hence there will be an 'optimum' LO power level which provides the best possible conversion gain for a particular mixer. Problems also arise if R_i is too high; for example, any real system will have non-zero capacitance. The system bandwidth may become restricted by the time constant that this produces with the resistance R_i.

The mechanisms which produce noise in bolometer systems also occur in heterodyne systems. In addition, heterodyne detectors suffer from a particular noise problem caused by the need to illuminate the mixer with LO power.

In general $P_l \gg P_s$ and the mean current flowing in the mixer will be essentially i_l. This LO-induced current will produce a shot noise level

$$\langle i_n \rangle^2 = 2ei_l B \qquad (7.38)$$

where B is the output measurement bandwidth (i.e. the range of frequencies that we observe being produced by the mixer).

Now the output IF signal power will be such that

$$\langle i_0 \rangle^2 = 2i_s i_l. \qquad (7.39)$$

Hence the output signal-to-noise power ratio, ignoring any other sources of noise, will be

$$\frac{S}{N} = \frac{i_s}{eB}. \qquad (7.40)$$

However,

$$i_s = \frac{e\eta}{h\nu} P_s \qquad (7.41)$$

which means that

$$\frac{S}{N} = \frac{\eta P_s}{h\nu B}. \qquad (7.42)$$

The factor η represents a quantum efficiency. This implies that $\eta \leq 1$. The best possible signal-to-noise power ratio will therefore be

$$\left(\frac{S}{N}\right)(\max) = \frac{P_s}{h\nu B}. \qquad (7.43)$$

A number of important conclusions may be drawn from this result. Firstly, it can be seen that the maximum possible signal-to-noise ratio does not depend explicitly upon the LO power level. This may initially be a surprise as it has been established that the conversion gain, and hence the output power level, tends to increase in proportion to the LO power. However, the noise level i_n also increases in proportion to the LO power. These two effects tend to cancel, producing a fixed signal-to-noise ratio.

In any real device the conversion gain will vary with the LO power level in a less simple way than we have assumed. There will also be a number of other processes generating noise within the receiver system. Hence the actual S/N ratio of a real system *will* peak at a particular LO power level. No real system, however, can be expected to provide a signal-to-noise performance in excess of the maximum value set by equation (7.43).

Secondly, it can be seen that the maximum possible S/N depends upon the signal and LO frequency ν. This may be seen as a consequence of the fact that each photon has an energy $h\nu$. The number of photons—and hence the current produced—by a given input power will therefore tend to fall if the frequency ν is increased. In this respect, high-frequency heterodyne systems cannot work as well as those used at lower radio frequencies.

Finally, S/N depends upon the output IF bandwidth B. The shot noise generated within the mixer has a uniform power spectrum. Hence, if we increase the bandwidth, then we shall observe more noise from the mixer.

In some cases the signal power P_s is at a single frequency or is confined to a narrow frequency range. Under these circumstances it is an advantage to use a mixer (or employ passband filters) which restricts B to the lowest value which passes the IF output produced by the signal.

The noise performance of a heterodyne system is often indicated by quoting a *noise temperature*. This is a useful method as it is a value which may be measured fairly easily and which relates directly to a common application of millimetre-wave heterodyne systems.

Figure 7.3(b) illustrates a practical mixer system being used to detect the output from a 'black-body' *thermal* source. The source is large enough to fill the beam pattern of the detector antenna completely. The radiation output from such a source is determined simply by its physical temperature T.

The *Wien displacement law* states that the peak output power per hertz bandwidth from a thermal source occurs at the wavelength λ_m such that

$$\lambda_m T = 5.1 \text{ mm K}. \tag{7.44}$$

This result is based upon locating the maximum of the *Planck curve* which describes the output spectrum of a black-body thermal source.

At signal wavelengths $\lambda \gg \lambda_m$, it can be shown from the Planck curve that the signal power in a bandwidth B is simply

$$P_s = kTB. \tag{7.45}$$

The portion of the spectrum which satisfies the above inequality is conventionally referred to as the *Rayleigh–Jeans region*. In almost all cases of practical interest it is found that a millimetre-wave receiver observing a thermal source is operating in the Rayleigh–Jeans region. A typical example is a system being used to make thermal maps of the earth from orbit. The temperature of the earth would be around 300 K, implying that $\lambda_m \sim 10 \,\mu\text{m}$. A heterodyne system observing the earth at a few millimetres wavelength would therefore be working well into the Rayleigh–Jeans region.

The mixer in figure 7.3(*b*) is assumed to have a conversion gain G and passes its output signal to an amplifier of power gain A. A bandpass filter is used to define the signal bandwidth B and the IF signal power level is measured using a meter whose output is averaged over the time τ.

Any noise generated in the mixer (or in the following amplifier) is assumed to come from a mythical thermal source of temperature T_n which injects unwanted power into the mixer. For a *single-sideband* receiver the mean output voltage provided by a power meter will be proportional to

$$V = kBGA(T + T_n). \tag{7.46}$$

By using two thermal sources of temperatures T_1 and T_2 we can obtain two output levels V_1 and V_2 such that

$$V_1 = kBGA(T_1 + T_n) \qquad V_2 = kBGA(T_2 + T_n). \tag{7.47}$$

By combining and rearranging these expressions we can show that

$$T_n = \frac{T_1 V_2 - T_2 V_1}{V_1 - V_2}. \tag{7.48}$$

The value T_n can therefore easily be measured given two thermal sources of known temperatures. Although it provides no information regarding the origins of the system noise, the *system* (or *receiver*) *noise temperature* T_n is a simple measure of the amount of noise.

If the mixer responds equally to both input signal sidebands, then the system acts as a double-sideband receiver. The signal power falling upon the mixer is now $2kTB$ rather than kTB and produces twice the IF output of a single-sideband system. This doubles the output voltage measured by the power meter. For an unchanged level of mixer–amplifier noise this halves the value of T_n obtained from equation (7.48).

To distinguish between these two cases it is necessary to specify whether the measurement includes power from both sidebands or not. This has produced the convention of specifying receiver noise temperatures as being either the *double-sideband* or *single-sideband* value. In practice the noise temperature of a double-sideband system is often measured using a thermal source and the value obtained is doubled before being quoted as a 'single-sideband' result. When comparing different systems, care must be taken to ensure that noise temperatures are compared on a 'like-with-like' basis to avoid confusion.

Where we are interested in the performance of a *mixer* rather that than of a whole system, it is conventional to quote a *mixer noise temperature* along with the conversion gain. Many of those working on mixers treat the mixer noise as a thermal source 'after' the mixer, i.e. defined so that the mixer noise power P_m presented to the following amplifier is

$$P_m = kT_m B \qquad (7.49)$$

where T_m is the mixer noise temperature. This definition has the advantage that it is the same for double- or single-sided systems. Confusion may arise, however, because this definition omits the conversion gain factor G which appears in the definition of system noise temperatures. Hence the value of T_m in a particular system will often be much lower than T_n.

When using a receiver the important parameter is the overall system noise. Details of the amounts of mixer noise, amplifier noise, etc, are important only when designing or modifying heterodyne systems.

The actual output measurement will be of the voltage produced by the IF power detector, averaged over a time τ. One of the properties of random ('white') noise is that a noise power whose mean power level is P_n will fluctuate from moment to moment by an amount proportional to $\sqrt{P_n}$. Hence the mean fluctuation in the measured output will vary proportionally to $1/\sqrt{B}$. The spectrum of these fluctuations is also uniform ('white') over the frequency range $0-B$ Hz.

The device which measured the IF power level is often called the detector. This usage is sometimes a little confusing, but it is reasonable because the device acts in the same way as a bolometric detector, i.e. it provides an output signal proportional to the overall input power level. The output time constant is then conventionally referred to as the *post-detection filter* to distinguish it from the IF filter.

The time constant acts as a low-pass filter, only passing fluctuations in the

Mixers and Heterodyne Detection

detected level which vary with frequencies below $1/2\pi\tau$. Hence the observed S/N ratio at the filter output will depend upon $1/\sqrt{\tau}$. Combining this with the effect of the IF filter we may conclude that the overall S/N behaviour of the system will allow us to detect a change in signal level corresponding to a thermal source of temperature

$$\Delta T = \frac{T_n}{\sqrt{B\tau}}. \qquad (7.50)$$

In practice, a real receiver will differ in detail from the simple system assumed here. This will mean that various correction factors should be included in the above expressions. The results obtained here are, however, a good guide to the general properties of heterodyne signal detection systems.

From the expressions given earlier for the power from a thermal source and the maximum possible S/N ratio performance it follows that the minimum possible system noise figure for a single-sideband receiver will be

$$T_n(\min) = \frac{h\nu}{k} \qquad (7.51)$$

i.e. if ν is in gigahertz and T_n is in kelvins

$$T_n = 0.047\nu. \qquad (7.52)$$

At a signal frequency of 300 GHz this implies that a single-sideband receiver cannot have a system noise temperature of less than 14 K.

At present, practical millimetre-wave heterodyne receivers have system noise temperatures which are much higher than this ideal limiting value. InSb bulk mixers are generally able to provide the lowest practical noise temperatures. Over the 100–500 GHz range, InSb receivers have been made which provide double-sideband system noise temperatures of around 200–300 K. The InSb mixer is able to offer good performance because it is cooled to a few kelvins, reducing the background thermal noise level. Being a bulk mixer it is also often better matched to the input signal power.

In comparison, Schottky diode mixers operating at room temperature generally offer system noise temperatures around 1000 K. This can be reduced to around 500 K by cooling the mixer diode to around 20 K. At lower temperatures the performance of Schottky devices deteriorates because of the electronic properties of the materials.

Although offering lower noise levels, InSb mixers are limited to an IF bandwidth of less than 1 MHz if cooled to obtain the lowest possible noise. A typical Schottky device is able to offer a bandwidth which is hundreds or thousands of times greater. The choice of mixer will therefore be affected by the bandwidth of the incoming signal power.

At present a considerable amount of attention is being devoted to SIS mixers. These offer wide IF bandwidths, comparable with or better than those of Schottky devices, but are able to operate when cooled to much lower temperatures. Heterodyne receivers based upon SIS mixers have already produced system noise temperatures below 100 K at 100 GHz. It is probable that this level of system noise will become available at higher frequencies over the next decade or so.

Chapter 8

Filters and Resonators

Many optical systems require the use of filters or frequency-selective arrangements. In most cases the filter has a fairly simple task, e.g. defining the range of signal frequencies coupled onto a bolometric detector. However, it is worth noting that, in conventional electronics and microwave systems, frequency-sensitive arrangements are used for a number of other functions. These include impedance matching, reflection suppression, and diplexing or multiplexing. All these functions may also be provided in optical systems by employing appropriate arrangements.

In this chapter we shall outline the properties of three general types of optical filter:

(i) semi-reflecting sheets;
(ii) Fabry–Perot and resonant filters;
(iii) absorption filters.

8.1 Semi-reflecting Sheets

One of the simplest types of frequency-selective filters to manufacture is a plane parallel-sided slab or sheet of dielectric material. Figure 8.1 represents a sheet of thickness d and refractive index μ placed in air. A signal beam incident upon the sheet will generate a reflected wave when it strikes each of the air–dielectric interfaces.

For simplicity we shall consider the example of an essentially plane parallel beam striking a thin sheet of loss-free dielectric at normal incidence. The presence of *two* semi-reflecting interfaces gives rise to interference effects. Hence the total reflectivity and transmissivity of such a sheet are frequency dependent, even when the refractive index—and hence the reflectivity of a single interface—is not.

Figure 8.1(a) illustrates the customary optical approach for calculating the overall reflectivity of such a sheet. An incident signal of amplitude A is incident upon the first air–dielectric interface. The reflectivity of this interface for a wave arriving from the air side will be

$$r = \frac{\mu - 1}{\mu + 1} \qquad (8.1)$$

and the transmissivity will be

$$\tau = 1 + r. \qquad (8.2)$$

The first contribution to the total reflected field will, therefore, be equal to Ar.

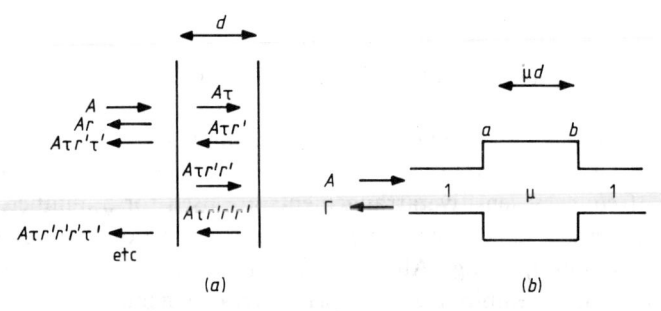

Figure 8.1 (a) Multiple reflections produced by field incident upon a parallel pair of semi-reflecting interfaces. (b) Reflection produced by an analogous length of mismatched transmission line.

The transmitted wave will propagate through the dielectric and strike the second interface whose reflectivity is

$$r' = \frac{1 - \mu}{1 + \mu} = -r. \qquad (8.3)$$

Hence the magnitude of the first reflection component from this second interface will be $A\tau r'$. This returns to the first interface where it causes a wave of magnitude $A\tau r' \tau'$ to pass out of the dielectric, contributing to the total reflected field. This contribution will differ in phase from the initial reflection by an amount 2δ, where

$$\delta = \frac{2\pi d\mu}{\lambda}. \qquad (8.4)$$

The total field produced by these first two reflection components will therefore be

$$\Delta R = Ar + A\tau r' \tau' \exp^2(-j\delta) \qquad (8.5)$$

where ΔR is the portion of the total reflected field R which is produced by these first two reflected components.

Now it can be seen that

$$\tau' = 1 + r' = 1 - r \tag{8.6}$$

and $r' = -r$; hence

$$\Delta R = Ar + A(1 - r^2)[-r \exp^2(-j\delta)]. \tag{8.7}$$

By continuing the process of summing the contributions produced following further reflections between the interfaces we can obtain the expression

$$R = Ar + \frac{A(1 - r^2)}{-r} \sum_{n=1}^{\infty} [-r \exp(-j\delta)]^{2n} \tag{8.8}$$

which is equivalent to

$$R = Ar + \frac{A(1 - r^2)[-r \exp^2(-j\delta)]}{1 - r^2 \exp^2(-j\delta)}. \tag{8.9}$$

If the overall reflectivity Γ is defined to be equal to R/A we can now say that

$$\Gamma = r + \frac{(1 - r^2)[-r \exp^2(-j\delta)]}{1 - r^2 \exp^2(-j\delta)}. \tag{8.10}$$

A similar method may be employed to calculate the total signal transmissivity of the dielectric sheet.

The value of r depends only upon the refractive index μ of the dielectric (relative to the bounding medium). The phase δ, however, depends upon d/λ. This means that the reflectivity of the sheet is wavelength (and frequency) dependent.

In fact, a much simpler method for calculating the sheet reflectivity can be used. This is based upon regarding the sheet as a length of transmission line which has a characteristic impedance differing from the 'air' transmission line which is attached to it on either side. This method is illustrated in figure 8.1(b).

The impedance of free space is normalized to unity and the characteristic impedance of the dielectric is taken to be $Z_c = \mu$. a represents the interface at the side upon which the input beam is incident.

For a wave passing through the dielectric and arriving at b the 'output' transmission line will appear as a termination load whose impedance is unity. Hence the signal reflectivity Γ_b at this interface will be

$$\Gamma_b = \frac{1 - \mu}{1 + \mu}. \tag{8.11}$$

The same reflectivity would be produced if an impedance

$$Z = Z_c \frac{1 + \Gamma_b \exp(-j2\delta)}{1 - \Gamma_b \exp(-j2\delta)} \tag{8.12}$$

where δ is as defined above were positioned at a, replacing the arrangement shown. The total reflectivity Γ of the system will be

$$\Gamma = \frac{Z-1}{Z+1}. \tag{8.13}$$

Combining these three expressions we arrive at the result

$$\Gamma = \frac{(\mu^2 - 1)[1 - \exp(-j2\delta)]}{(1+\mu)^2 - (1-\mu)^2 \exp(-j2\delta)}. \tag{8.14}$$

At first glance this result appears different from that obtained using the 'optical' multiple reflection method. However, equation (8.10) may be rearranged into the form

$$\Gamma = \frac{r - r\exp(-j2\delta)}{1 - r^2 \exp(-j2\delta)}. \tag{8.15}$$

Using equation (8.1) to substitute for r, we obtain

$$\Gamma = \frac{\mu - 1}{\mu + 1} \frac{1 - \exp(-j2\delta)}{1 - (\mu - 1)^2 \exp(-j2\delta)/(\mu + 1)^2} \tag{8.16}$$

which may be seen to be identical with equation (8.14) if we multiply through by $(\mu + 1)$.

The behaviour of a dielectric sheet may therefore be analysed using either an 'optical' approach or an 'electronic' approach based upon the concepts of impedance and transmission lines. In this case, as in many others, the transmission line–impedance model makes the actual analysis simpler to perform. For example, here it enabled us to avoid having to know the solution of the infinite series produced in equation (8.8).

Dielectric sheets are quite simple to manufacture and use. They do, however, suffer from a couple of drawbacks. Firstly, the maximum possible reflectivity of a single sheet is limited by the value of μ. The reflectivity reaches a maximum value of

$$\Gamma(\max) = \frac{\mu^2 - 1}{\mu^2 + 1} \tag{8.17}$$

when $\exp(-j2\delta) = -1$.

The refractive indexes of some typical millimetre-wave dielectrics and the basic behaviour of a dielectric slab have already been mentioned in chapter 2. HDPE has a refractive index of about 1.53. Hence a slab of HDPE would have a peak signal reflectivity of around 0.39. This is equivalent to a power reflectivity $|\Gamma|^2$ of about 0.16.

In order to obtain a high peak reflectivity a higher dielectric constant is required. Mylar and Kapton may be obtained in the form of rolls of thin sheet. Both of these materials have a higher loss tangent than HDPE but have refractive indexes μ of around 1.75 which permits a peak signal reflectivity of around 0.50 (i.e. $|\Gamma|^2 \simeq 0.25$).

Filters and Resonators

Quartz ($\mu \simeq 2.1$; $|\Gamma|^2 \simeq 0.40$) and sapphire ($\mu \simeq 3.1\text{--}3.4$; $|\Gamma|^2 \simeq 0.66\text{--}0.70$) can provide higher peak reflectivities. These materials are, however, more expensive and fragile and may be polarization sensitive. Large thin sheets of these materials are almost impossible to fabricate.

The reflectivity of a dielectric sheet varies periodically with changes in signal frequency. This is useful in some cases but may be a problem when a reflector is required which has a uniform reflectivity over a given frequency range. Also, the phase of Γ varies with frequency in a complex way. This may cause problems in arrangements where phase is significant.

8.2 Metallic Mesh

Many millimetre-wave systems employ an alternative form of semi-reflecting element, the *metallic mesh* (or *grid*). This has a number of advantages over a dielectric sheet. In particular, it is possible to design and produce metallic meshes which have almost any reflectivity in the range $|\Gamma| = 0\text{--}1$. It is also often possible to design a mesh whose reflectivity varies with signal frequency in a particular way.

Various types of metallic mesh may be used as semi-reflectors. Figure 8.2 illustrates some of the most common forms. Each mesh consists of a repetitive pattern of metal shapes arranged in a plane. In some cases the metal is continuous and hence may be formed by cutting or etching shaped holes into a thin metal sheet. In other cases each metal shape is disconnected from its neighbours and the shapes must then be supported by a dielectric substrate. Patterns made without a supporting dielectric are called *free-standing* meshes.

Figure 8.2(*a*) illustrates a typical mesh pattern of metal squares separated by narrow gaps. The reflection properties of such a mesh depend upon the gap width g and the mesh *period* p compared with the wavelength λ of the incident wave. A detailed analysis of the reflection behaviour of a metal mesh is quite complex. We can, however, establish the basic frequency-sensitive properties of a mesh by considering how a metal surface reflects an incident wave.

Consider just one of the metal squares shown in figure 8.2(*a*) illuminated with a signal of wavelength $\lambda \ll p - g$. The incident field sets up an oscillatory current in the metal. Over most of the metal surface this current is uniform. This means that for any small area, well away from the edges of the square, any charge flowing into the area from one direction is exactly equal to the amount leaving in the opposite direction. This uniform current flow produces a reflected wave much as in the case of an ordinary mirror.

However, charge cannot flow across the edges of the square. The current at the edges of the square must always be zero when measured in the direction perpendicular to the edge. In fact, oscillatory charge accumulations occur in the region which is a distance comparable with λ from the edge. This charge impedes the current flow, reducing its magnitude locally and altering its oscillatory phase compared with the incident wave.

When $\lambda \ll p - g$, the current distribution over most of the square is the same as if it had been a plane surface infinite in size. The square hence produces a reflected wave just as if it were a metal mirror. However, when $\lambda \simeq p - g$, the current induced over the whole square becomes affected by the presence of the edges. When $\lambda \gg p - g$, the incident wave is virtually unable to set up a current flow as any small current swiftly causes a charge distribution to appear on the square whose potential field opposes any further flow.

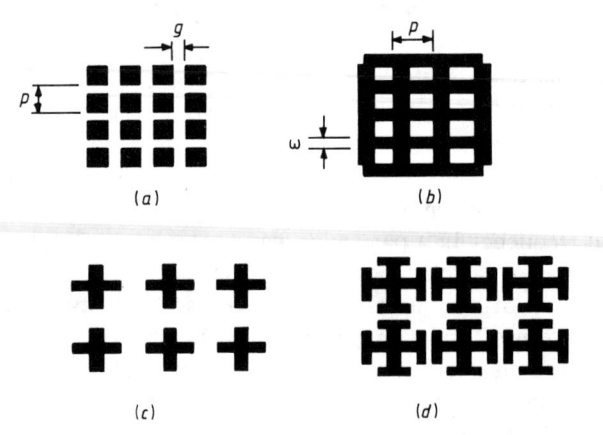

Figure 8.2 Examples of some periodic metallic mesh patterns with different types of frequency-dependent reflectivity. (a) Capacitive mesh. (b) Inductive mesh. (c) Resonant crosses. (d) Jerusalem crosses.

A pattern of such squares behaves virtually as a metal mirror at signal wavelengths which are much smaller than the square size (or the period p which is almost identical with the square size). At lower frequencies, when the signal wavelength is much larger than the mesh period, an incident wave sets up almost no current. In the absence of any induced current flow the wave propagates through the mesh without taking much notice of its presence.

The properties of such a mesh can therefore be represented as a *capacitance* shunting a transmission line along which the signal beam is assumed to propagate. At low frequencies the capacitance behaves as a high impedance and has little effect upon the passing signal. At high frequencies the capacitance behaves as a low impedance and acts almost like a short circuit, reflecting the incident signal. A mesh of the form shown in figure 8.2(a) is therefore often referred to as a *capacitive* mesh or grid.

Filters and Resonators

The mesh shown in figure 8.2(b) is the 'inverse' of a set of metal squares. It is therefore, perhaps, not surprising to discover that its reflection properties are essentially the inverse of a capacitive mesh and it is often referred to as an *inductive* mesh. This form of mesh can be regarded as being composed of two perpendicular arrays of parallel wires.

Any incident field can be represented as being the sum of two orthogonally polarized components whose electric vectors are parallel to the two wire directions. In general we may assume that the width (or diameter) w of each wire is much smaller than the signal wavelength. Any electric field vector component which is perpendicular to a particular wire's axis will be unable to induce any significant current 'across' the wire. However, any field component parallel to the wire will be able to induce a current along the wire.

For a signal wavelength $\lambda \gg p - w$, the array of wires acts almost as if it were a continuous metal sheet. The incident wave sets up currents along the wires which produce a reflected wave and the signal power is reflected as if the mesh were a plane mirror.

When $\lambda \ll p - w$, field can pass through the 'gaps between the wires'. This is because any particular part of the incident wave can only induce a significant current when it passes sufficiently close (i.e. within a wavelength or so) to a parallel wire. Hence, if $\lambda \ll p$ and $w \ll p$, we find that almost all the incident field will pass through the mesh as if it did not exist.

The behaviour of this form of mesh can therefore be represented in terms of an inductance shunting a transmission line along which the signal beam is assumed to propagate. At low frequencies the inductance behaves as a short circuit, reflecting the signal. At high frequencies the inductance behaves as a very high impedance and has almost no effect upon the signal transmission.

In practice, the behaviour of these two forms of metal mesh is more complex than has been described. For example, when the dimensions of the squares (or square holes) become comparable with $\frac{1}{2}\lambda$ a *resonant* current pattern will be created. This may cause the reflectivity to be higher than expected at this frequency. The need to support a capacitive array of squares with a dielectric film will modify the field properties near the mesh, altering its reflective properties. For many purposes, however, we can regard these grids as simple inductors or capacitors which produce a power reflectivity of about 0.5 when their period, $p \simeq \frac{1}{2}\lambda$.

A single capacitive or inductive mesh can be used as simple 'low-pass' or 'high-pass' filters. They can also be used over a moderate frequency range as a semi-reflector whose reflectivity may be set to any desired value by a suitable choice of p, g or w. For some applications, however, we require a reflector whose reflectivity is very high (or very low) over a limited signal frequency range. Such a mesh could then be used to select or reject a narrow band of signal frequencies.

Figure 8.2(c) illustrates a possible form of *resonant* mesh based upon an array of metal crosses. The reflective behaviour of these crosses can be explained by regarding them as a pair of crossed 'dipoles'. An incident wave will induce an oscillatory current along the dipole arms. The induced current will have a resonant maximum when the radiation wavelength λ is such that $l \simeq \frac{1}{2}\lambda$ if l is the length of the dipole arms.

An array of crosses hence acts as a reflector whose reflectivity is, in general, low but rises to a resonant peak at a particular frequency. In practice, plain cross arrays usually have a high reflectivity over a fairly wide frequency range. Various modifications to this basic pattern have been employed to obtain narrower ranges of the high reflectivity. A typical example is the *Jerusalem cross* pattern shown in figure 8.2(*d*).

The bars at the ends of the arms of a Jerusalem cross act as a reservoir for the charge flowing to the ends of the arms. They also increase the capacitive coupling between the charges at the ends of adjacent crosses. This tends to lower the resonant frequency slightly. When the input signal is exactly at the resonant frequency, the weak coupling between adjacent crosses tends to cause their resonances to act together, increasing the peak reflectivity. At frequencies just off the resonant peak the fields in adjacent crosses tend to be out of phase and the net effect is to reduce the mesh reflectivity. Hence the Jerusalem cross can provide a narrower resonant reflection band with a higher peak reflectivity.

The bars at the ends of the arms should not, however, be too large or too close to one another as this tends to 'over-couple' adjacent resonators. The behaviour of each resonator then becomes swamped in the ability of the overall metal pattern to behave like a metal sheet. The reflectivity of the array then becomes similar to a capacitive mesh of similar period.

A number of other array patterns may be used, e.g. patterns of small discs or squares with significant gaps. These tend to act as a combination of a cross resonator and a capacitive mesh. We may also use the 'inverse' of any pattern (replacing metal with gap and vice versa) as in the case of swapping metal squares (capacitive) for square holes in a metal sheet (inductive). In general this produces a mesh whose reflectivity is the reciprocal of its inverse.

Each of these thin meshes can be treated as a shunt impedance across a transmission line. One interesting result which arises from this is that, for a loss-free mesh, the magnitude and phase of the reflected wave are related quite simply. This follows from the general result obtained in chapter 5 that, for a loss-free shunt on a transmission line, the signal reflectivity Γ and transmissivity τ must satisfy the two expressions

$$\tau = 1 + \Gamma \qquad 1 = |\tau|^2 + |\Gamma|^2 \qquad (8.18)$$

in order that the field be continuous at the same-reflector and that the output and input powers should be the same.

As a consequence we can expect that τ and Γ will differ in phase by $90°$

Filters and Resonators

and that the phase change θ upon reflection will be such that

$$|\Gamma| = |\cos\theta|. \tag{8.19}$$

The presence of a dielectric supporting sheet will modify this result. However, provided that the sheet is very thin compared with the wavelength and its dielectric constant is not greatly different from unity, the change that this produces will be slight.

8.3 Fabry–Perot and Resonant Filters

Single semi-reflecting sheets are fairly easy to manufacture and simple to use. Metal patterns can be formed by photographic and chemical etching or vacuum deposition techniques with small periodic structures. However, they suffer from some limitations which make them unsuitable for certain applications. For example, we may require an arrangement whose frequency-selective transmissivity or reflectivity can be altered while in use. This is effectively impossible using a single mesh whose properties are essentially fixed at the point of manufacture.

Difficulties also arise when we require a bandpass or band-reject filter which acts over a very narrow frequency range. Highly resonant grids are difficult to make and—once made—we find that the mesh differs slightly from what was required. It must then be thrown away and another made. Similarly, the properties of such a mesh may prove sensitive to changes in temperature or some other quantity outside our control.

One of the most common forms of adjustable optical filter is the *Fabry–Perot resonator*. This normally consists of two semi-reflecting sheets placed a particular distance d apart. As in the case of the dielectric sheet discussed earlier, an incident wave generates multiple reflections from the two semi-reflectors. Interference between the reflected components produces an overall reflectivity and transmissivity which can be strongly frequency dependent. The resonant peaks and dips in the frequency response of the resonator can then be 'tuned' to different frequencies by adjusting the spacing d between the semi-reflectors.

It is customary in optics texts to analyse the Fabry–Perot resonator in terms of plane-wave beams. This leads to results which are broadly similar to those obtained earlier for a parallel-sided dielectric slab. However, we find that, at millimetre-wave frequencies, the effects of beam diffraction often have a significant effect upon the performance of a Fabry–Perot resonator. In order to take diffraction into account we can employ a Gaussian beam approach.

Figure 8.3 illustrates a Gaussian beam incident upon a Fabry–Perot resonator which is composed of two plane parallel semi-reflectors positioned normal to the beam axis.

The phase-front radius of curvature and size are not changed when the field is reflected; hence we can think of the field inside the resonator as being a 'folded-up' version of the beam which would have existed if the resonator were removed. Each reflection will introduce a phase change and reduce the amplitude of the reflected field without changing its overall pattern.

If the semi-reflectors had no effect, then an input beam of field ψ whose beam waist plane is a distance Z from the second semi-reflector would generate a field $\psi(Z)$ at the plane of the second semi-reflector. For simplicity let us consider the case where the two semi-reflectors are identical and loss free. The first contribution to the total transmitted field Ψ produced by the input beam will then be

$$\Psi_1 = t^2 \psi(Z) \tag{8.20}$$

where the subscript 1 indicates that this is just the first contribution and t is the signal transmissivity of a single semi-reflector.

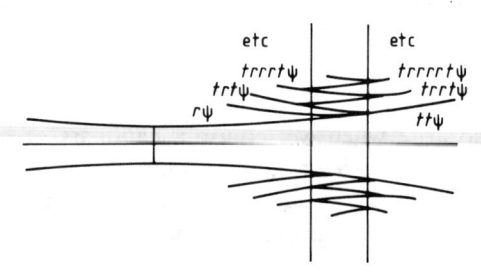

Figure 8.3 Effect upon a Gaussian beam incident upon a parallel pair of semi-reflectors (i.e. a plane Fabry–Perot resonator).

The second contribution, which results from a single reflection in each of the semi-reflectors, produces an output

$$\Psi_2 = t^2 r^2 \psi(Z + 2d) \tag{8.21}$$

where d is the distance between the two semi-reflectors and r is the signal reflectivity of a single semi-reflector.

Summing the contributions produced by repeated reflections we obtain the total output field

$$\Psi = t^2 \sum_{n=0}^{\infty} r^{2n} \psi(Z + 2nd). \tag{8.22}$$

A similar argument can be applied to define the total field reflected by the resonator.

Now the field Ψ pattern will differ from ψ; hence the power emerging from the resonator will not couple with perfect efficiency into a sensor whose antenna pattern has been designed to couple a beam of the form ψ perfectly. When calculating the amount of signal power coupled through the

Filters and Resonators

resonator, we must therefore distinguish between the *total* power transmitted and that part which will couple into a given output beam.

The total output power P_t transmitted by the resonator will be

$$P_t = \langle \Psi | \Psi \rangle \qquad (8.23)$$

but the power coupling into a given output beam of field Φ will be

$$N = \frac{|\langle \Psi | \Phi \rangle|^2}{\langle \Phi | \Phi \rangle}. \qquad (8.24)$$

In most practical cases we are concerned with the amount of power coupling when the input is a fundamental-mode Gaussian beam and the output beam is essentially identical with the beam which would have emerged if the resonator had no effect upon the beam pattern. Provided that ψ is correctly normalized to unity total beam power this means that we can write

$$N_0 = |\langle \Psi | \psi_0(z) \rangle|^2. \qquad (8.25)$$

By using the methods outlined in chapters 1–4 this may be seen to be equivalent to

$$N_0 = |C_0|^2 \qquad (8.26)$$

where

$$C_0 = t^2 \sum_{n=0}^{\infty} \frac{r^{2n} \exp(j2nd2\pi/\lambda)}{1 - jnd\lambda/\pi\omega_0^2}. \qquad (8.27)$$

Now, provided that r^{2n} approaches zero when $nd\lambda/\pi\omega_0^2$ approaches unity, this expression is essentially the same as if the beam were a plane parallel one which propagated without diffraction. More generally, however, we find that the power coupling N_0 is modified by the Gaussian nature of the beam.

One interesting conclusion of the above analysis may be drawn from noting that Z, the distance to the input beam waist plane, does not appear in equation (8.27). This means that an alteration of N_0 caused by Gaussian diffraction is independent of the position of the resonator along the beam. Despite this, it is usually advisable to position the resonator near the beam waist. This is because the above analysis ignores any effects of beam truncation which may arise when real semi-reflectors of finite aperture size are used. The beam is at its smallest near the beam waist; hence truncation effects are minimized when the resonator is located near this plane.

Figure 8.4 illustrates a calculation of how the power transmissivity of a typical Fabry–Perot instrument varies with frequency. The example shown is for a pair of loss-free capacitive meshes, each having a signal reflectivity $|r|$ of 0.85 at a chosen 'nominal design' frequency of 100 GHz. Two power transmissivity curves are plotted. The curve which reaches a peak value of unity is calculated on the assumption that the beam can be regarded as simply plane parallel. The curve which peaks at a lower value takes into account the effects of diffraction for a pure fundamental-mode Gaussian beam. This calculation also includes any power loss which occurs because the output beam is no longer a fundamental-mode Gaussian of the type directed onto the resonator from the source.

The curves shown are for a resonator whose semi-reflectors are separated by 3.25 wavelengths at 100 GHz (i.e. 9.74 mm). If we ignore any phase shift produced when a beam is reflected by either semi-reflector, then for an identical pair of semi-reflectors we would expect transmission maxima to occur when the signal wavelength λ is related to the distance d between the semi-reflectors by the expression

$$\lambda(\tfrac{1}{2}m + \tfrac{1}{4}) = d \qquad (8.28)$$

where m is an integer. (Note that this differs from a dielectric slab whose interface reflectivities are opposite in sign.)

The phase shifts which take place each time that a field is reflected by one of the meshes alter the frequency at which the peak transmissivity occurs. A capacitive mesh slightly delays the phase of a reflected signal. This causes a signal to take slightly longer in passing back and forth between the semi-reflectors, as if they were further apart. Hence the peak transmissivity is shifted down to a lower frequency (i.e. a longer wavelength) than expected. In the case illustrated, the peak is shifted down from the 'design' value of 100 GHz to around 95 GHz. Inductive meshes would tend to shift the peak to a frequency which was a similar amount *above* the nominal value.

Many millimetre-wave Fabry–Perot resonators are made using a combination of one inductive and one capacitive mesh. This is for a number of reasons; for example, at very high or low frequencies one mesh becomes transparent and the other highly reflective, hence suppressing transmission over a wide frequency range well away from the desired resonance. One particular advantage is that the phase shifts of the two meshes tend to cancel. This results in the occurrence of peak transmissivity almost at the wavelength predicted by equation (8.28).

The lower curve shown in figure 8.4 is for a fundamental Gaussian mode whose beam waist size ω_0 equals 10 mm. This causes the peak power transmissivity, into an output beam whose form matches that of the initial input, to fall to 0.899. The frequency of peak transmission is also shifted slightly from the 'plane parallel' result. This is because the rate of phase change along a Gaussian mode differs from that of a plane wave.

Knowing that, for the effects of beam diffraction to be negligible, we would require that $r^{2n} \simeq 0$ when $nd\lambda/\pi\omega_0^2 \simeq 1$, we can introduce a 'figure of merit',

$$M = 1 - r^b \tag{8.29}$$

where

$$b = \frac{2\pi\omega_0^2}{\lambda d} \tag{8.30}$$

i.e. b corresponds to the value of n which would cause $nd\lambda/\pi\omega_0^2$ to equal unity.

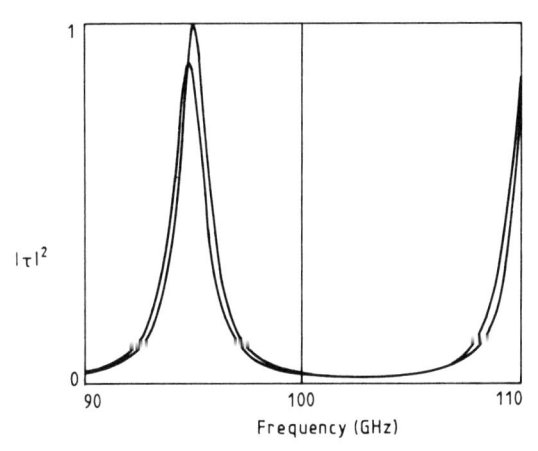

Figure 8.4 Plots of the calculated reflectivity versus signal frequency for a typical plane Fabry–Perot resonator formed using capacitive meshes. The curve which peaks at unity reflectivity assumes a 'plane' input field. The curve which peaks below unity is for a fundamental Gaussian beam. $\omega_0 = 10$ mm, semi-reflector separation = 3.25 wavelengths; $P_M = 0.898$; $F_M = 94.800$; figure of merit, 0.97.

This value is a convenient way of estimating the effects of beam diffraction as we find that, over a wide range of values, M is roughly equal to the peak transmissivity produced when diffraction is taken into account.

Fabry–Perot resonators are relatively simple to design and can be tuned in use without difficulty simply by altering the spacing of the semi-reflectors. Their performance, however, becomes dominated by diffraction problems when high levels of performance are required. In some cases an improved filter function can be obtained by using a number of meshes of different types at varying spacings to build a 'multi-element' filter. Arrangements of this type can be analysed using the analogy of a set of shunt loads spaced out along a transmission line. Filters using up to a dozen or so meshes have been used successfully. A drawback of these arrangements is that they are almost impossible to adjust in use. They do, however, make excellent 'fixed' filters.

In order to narrow the width of the transmission peak of a Fabry–Perot resonator we must increase either the reflectivity r or the spacing d. This will reduce the value of M unless we also increase the beam waist size ω_0 by an appropriate amount. This may mean that, in order to avoid truncation problems, the Fabry–Perot resonator may need to be rather large in diameter.

Many practical systems also require filters which can be placed at an angle to the incident signal beam. This leads to the diffraction problem called *walk-off* which is illustrated in figure 8.5(*a*).

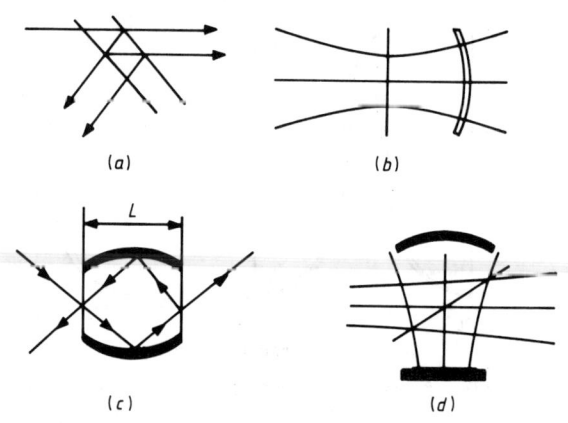

Figure 8.5 The problem of 'walk-off' caused by non-normal incidence (*a*), and some possible arrangements which avoid this problem: (*b*) semi-reflectors curved to match beam phase front; (*c*) 'ring' resonator of two curved mirrors and two plane semi-reflectors; (*d*) matched resonator.

The reflected field passing back and forth between the two semi-reflectors is displaced sideways as a consequence of striking the meshes at an angle to the normal direction. This means that the various field components emerging from the resonator are displaced sideways and no longer share the same beam axis. As a result, the components do not interfere as intended and the output beams are difficult to couple efficiently through an optical system into a detector. Any resonant peaks in transmissivity or reflectivity will be degraded in much the same way as if M (the Gaussian 'figure of merit') were too small.

The deterioration in performance which occurs when M is small is also partly a result of field being displaced perpendicular to the normal direction as it reflects back and forth. Hence the problems which arise if ω_0 is too small are also referred to as *walk-off*.

Filters and Resonators

8.4 Refocusing and 'Ring' Resonators

Various modified forms of resonator have been devised in order to maintain good performance for a compact arrangement where ω_0 is relatively small. These resonators employ one or more *curved* reflectors arranged so as to correct for the effects of beam diffraction.

Figure 8.5(*b*) illustrates the basic argument behind this method. Two reflectors are placed along a Gaussian beam. The reflectors have spherical curvatures R and R' which are identical with the phase-front curvature of the beam at the two locations chosen. Hence the field everywhere is incident locally normal to the surfaces.

A field moving in such a beam between the reflectors will hence be reflected back and forth, each reflected beam having the same beam waist size and location. If the reflectors are, in fact, semi-reflecting meshes, this means that the curvature performs a corrective focusing action. Now each emerging field component will be identical in form with the initial input beam. Only its overall phase and magnitude will be changed by the repeated reflections.

If we input, say, a fundamental Gaussian mode, then the output will be of this same form. Each trip of the reflected field back and forth and across the resonator will produce the *same* phase shift; hence it becomes possible in principle to obtain total transmissivities or reflectivities of unity (or zero). The resonator performance is no longer limited by the beam waist size ω_0 or a figure of merit.

In practice it is difficult to produce curved semi-reflecting meshes. It is more convenient to separate the tasks of semi-reflection and beam refocusing. Figure 8.5(*c*) illustrates a typical practical system called a *ring* resonator. Here the beam is passed around inside a 'square box'. Two walls are formed by a pair of plane meshes. One or both of the remaining walls is curved so as to refocus the reflected beam back into the form of the initial input.

Ring resonators of this type work quite well, even when the beam waist size is just a few wavelengths. Hence they provide compact frequency filters. The resonant frequency may be tuned quite easily by moving one or more of the reflecting walls, changing the size of the box. Another advantage of the ring resonator is that it functions in a way which easily permits the reflected power to emerge in a direction at right angles to the input. This is useful where the reflected output power is required.

A limitation of the performance eventually arises owing to the curved refocusing mirrors being used 'off axis'. When the beam waist size is small, the beam curvature alters rapidly along the propagation axis. This means that the mirror surfaces need to focus more strongly. Beam distortions may then become significant and the refocused field is no longer indistinguishable from the input beam.

An alternative arrangement is shown in figure 8.5(*d*). Here just a single semi-reflector is used to couple in and out of a resonator formed by two mirrors. This system acts in essentially the same way as shown in figure 8.5(*a*) with the single mesh providing a way in and out of the resonator.

The behaviour of these 'compact' systems which refocus the beam depends upon the properties of the input beam. If the input beam has a beam waist size or location which differs from that expected, then the refocusing action cannot perfectly correct the beam diffraction. Similarly, although the variations in phase-front curvature R and beam size ω of a Gaussian mode do not depend upon the mode numbers, the rate of phase change along the beam axis *does* depend upon the mode because of the anomalous phase term in the expression for a Gaussian mode. Hence the signal frequency at which a resonant peak in transmissivity or reflectivity occurs depends upon the beam mode numbers.

As the form of resonator shown in figure 8.5(*d*) employs an on-axis curved focusing mirror, it can be expected to be relatively immune from the beam distortions which arise when off-axis mirrors are used—as in the ring resonator illustrated in figure 8.5(*c*). However, ring resonators are particularly useful for a range of applications where access is required to both the transmitted and the reflected signals. For this reason they have proved more popular than the form shown in figure 8.5(*d*). Hence it is useful to analyse the properties of ring resonators. The main properties of the other arrangement can then be seen to be broadly similar.

For simplicity let us assume that it is possible to construct a resonator with arbitrary precision, that the effects of beam distortion due to the off-axis mirror are negligible and that the resistive dissipation losses of the metal meshes are zero. We can then concentrate on those problems which arise owing to the properties of the beams coupled through the resonator.

A ring resonator can be regarded as being a cube, of side size L, formed out of two almost plane focusing mirrors and two semi-reflectors. The axial path length d that the beam travels between focusing elements (or between semi-reflectors) will then be equal to $L\sqrt{2}$.

An input beam may be injected in the form of a single-mode Gaussian beam whose waist, of radius ω_d, is at the input semi-reflector. The profiles of the two focusing elements are chosen such that, at a specific *design frequency* f_d, this beam is refocused to form another beam waist, of the same size, at the other semi-reflector. Under these conditions, any beam power which is repeatedly reflected around inside the ring will always be refocused to have a field profile identical with the initial input. This situation is analogous to a simple Fabry–Perot resonator operating on a beam which does not suffer from beam diffraction. It differs, however, from a simple 'plane-wave' case in that the change δ in beam phase during a single traverse between the semi-reflectors will be

Filters and Resonators

$$\delta(m, n) = \frac{2\pi\sqrt{2L}}{\lambda_d} + 2(m + n + 1) \arctan\left(\frac{\lambda_d L}{\sqrt{2}\pi\omega_d^2}\right) \quad (8.31)$$

where m and n are the mode numbers of the beam mode, and $\lambda_d = c/f_d$.

For simplicity, we can assume that the semi-reflectors are a matched pair of an inductive and capacitive mesh, i.e. although having the same magnitude of signal reflectivity they produce opposing phase shifts upon reflection. The overall signal reflectivity Γ of the ring may now be shown to be

$$\Gamma(m, n) = \frac{r\{1 - \exp[-j2\delta(m, n)]\}}{1 - r^2 \exp[-j2\delta(m, n)]} \quad (8.32)$$

i.e. the reflectivity for a particular mode is similar to that for a plane wave striking a loss-free dielectric slab. The only difference is that the phase change δ per traverse is given by equation (8.31) instead of simply being $2\pi d/\lambda$.

Any resonant field patterns set up within the ring can now be characterized by the requirement that their phase-front curvature R should be identical at the focusing surfaces with

$$R_d = \frac{L}{\sqrt{2}}\left[1 + \left(\frac{\sqrt{2}\pi\omega_d^2}{\lambda_d L}\right)^2\right] \quad (8.33)$$

and must have a beam waist at each semi-reflector surface an axial distance $L/\sqrt{2}$ from the focusing surfaces.

Now for any Gaussian beam the beam size ω and phase-front radius R at a distance z from the waist will be such that

$$\frac{\pi\omega^2}{\lambda R} = \frac{\lambda z}{\pi\omega_0^2} \quad (8.34)$$

and we can also define the beam waist size using

$$\omega_0^2 = \frac{\omega^2}{1 + (\pi\omega^2/\lambda R)^2}. \quad (8.35)$$

Combining equations (8.34) and (8.35) we can obtain an expression for ω_0 which does not contain ω. Replace z with $L/\sqrt{2}$, R with R_d, and rearranging the result we obtain

$$\omega_0^4(\lambda) = \frac{\lambda^2 L(R_d - L/\sqrt{2})}{\sqrt{2}\pi^2} \quad (8.36)$$

to define the beam waist size $\omega_0(\lambda)$ which would ideally match the resonant field pattern within the ring resonator for a signal wavelength λ.

Using equation (8.33) to substitute for R_d we obtain

$$\frac{\omega_0^2(\lambda)}{\omega_0^2} = \frac{\lambda}{\lambda_d}. \quad (8.37)$$

Provided that we know the beam waist sizes of the beams coupled into and out of the ring we can use this result to calculate the variations in coupling efficiency with signal frequency.

It is of some interest to note that equation (8.37) does not depend upon the chosen ring size L. At first sight this may appear to imply that the ring may be made arbitrarily compact without affecting performance. This is not the case, however, as truncation and diffraction problems tend to produce losses whose severity increases as the ring size is reduced.

In order to limit the effects of truncation, the beam size ω_0 at the semi-reflectors (and ω at the focusing surfaces) must be sufficiently small compared with L. Furthermore, in order to minimize the effects of beam distortion caused by the use of off-axis reflectors, we must ensure that their curvature radii R_d is large. If we assume that the beam waist size at the semi-reflectors is such that

$$\omega_0 = \beta L \tag{8.38}$$

then we can expect that the beam size at the focusing elements, and their required radii of curvature will be such that

$$\omega^2 = \beta^2 L^2 + \frac{\lambda^2}{2\pi^2 \beta^2} \tag{8.39}$$

and

$$R_d = \frac{L}{\sqrt{2}} \left(1 + \frac{2\pi^2 \beta^4 L^3}{\lambda_d^2}\right). \tag{8.40}$$

Hence we can see that, as L falls towards zero, ω approaches a limiting value of $\lambda/\sqrt{2}\pi\beta$, and R_d also approaches zero. Hence there will unavoidably be a minimum value of L which can provide a required level of performance at a specific signal frequency.

Provided that we have a single-mode input beam whose beam waist size is equal to $\omega_0(\lambda)$, located at the input semi-reflector then the reflectivity of the ring may simply be calculated by employing equations (8.31) and (8.32) with the appropriate values of m and n. More generally, there will be a tendency for the actual input beam waist size ω_i to differ from ω_0 for signal wavelengths λ which differ from the design value λ_d. The behaviour of the resonator under these conditions is more complex, even when the incident beam remains single mode. This is because a given set of modes are mutually orthogonal only when they share a common beam waist size.

Consider the simplest case—that of a pure fundamental-mode ($m = n = 0$) input beam whose beam waist size differs slightly from the required value. By extending the techniques introduced earlier in this book and in some of the references we can now say that the reflected field Ψ will be of the form

$$\Psi = \sum \sum C\{0, 0, m, n\} \Gamma(m, n) \psi_{mn}(\omega_0). \tag{8.41}$$

Filters and Resonators

Here $\psi_{mn}(\omega_0)$ represents the normalized field pattern of the beam which would ideally couple into the (m, n)th mode within the resonator. $C\{0, 0, m, n\}$ represents the amplitude coupling coefficient:

$$C\{0, 0, m, n\} = \langle \psi_{00}(\omega_i) | \psi_{mn}(\omega_0) \rangle \qquad (8.42)$$

between the input beam and the (m, n)th mode. The precise behaviour of the system will now depend upon the variations in coupling and in the properties of the various modes. The amount of reflected power which is detected will also depend upon how efficiently the reflected beam of field Ψ matches the antenna or beam pattern of the detector. It should be noted that, although the input consisted of a single mode, the output will consist of a series of mode contributions.

The above argument is based upon a consideration of the reflectivity of the ring resonator. A simple analysis can also be undertaken to establish the signal transmissivity as a function of the various parameters. In general we can expect that the best level of performance will arise when the input and output beams are essentially single mode (preferably the fundamental mode) and are well matched to the resonator ω_0 value.

8.5 Absorption Filters

A number of materials have proved useful as millimetre-wave absorption filters. Most of these may be divided into two types: 'lossy' and 'loaded' dielectrics. In general the absorptivity of these materials increases with increasing signal frequency; hence they are suitable for making 'low-pass' filters.

Polymethyl methacrylate (PMMA) (or Perspex) and polyamide (nylon) are widely available polymers which can easily be made into useful absorption filters. Each has a refractive index which is almost frequency independent over the 100–500 GHz region. PMMA has a typical refractive index μ of 1.605 and, for polyamide, $\mu = 1.725$. Table 8.1 indicates how the absorptivity of these two materials varies with frequency.

As is the case with the low-loss polymers used for lenses, the absorptivity of these materials may vary significantly from sample to sample. In each case, however, the absorptivity rises with increasing signal frequency.

'Loaded' absorbers consist of a medium in which a number of small pieces of a different material are suspended. Two common examples are 'black polythene' and Flurogold (a tradename of Flurocarbon Inc).

Thin sheets of polyethylene (or other polymers) are often rendered black and opaque by mixing fine carbon grains (e.g. 'soot') into the material during manufacture. Common examples are the black 'dustbin bags' widely used for refuse collection in the UK. These blackened sheets often transmit millimetre wavelengths with almost no loss but strongly attenuate near-infrared and visible radiation.

Flurogold consists of a PTFE material containing grains of glass fibre.

A small particle will tend to scatter some of the radiation incident upon it. For particles which are smaller than the signal wavelength the amount of scattering will vary roughly in proportion with $1/\lambda^2$. Hence the transmission losses of a material loaded with small reflective particles will rise swiftly with increasing signal frequency. This rapid change in transmission loss with frequency may also be enhanced if the materials have dielectric properties which also vary significantly with frequency.

Table 8.1 Variation in absorptivity with frequency.

Frequency (GHz)	Absorptivity ($dB\,cm^{-1}$)	
	PMMA	Polyamide
150	1.9	1.5
200	3.0	3.8
250	4.4	5.4
300	6.3	7.7
350	8.7	10.7
400	11.4	13.6
450	13.9	17.8
500	16.7	21.2

For a typical sample of Flurogold the total transmission losses (including both absorption and scattering) will be around $5\,dB\,cm^{-1}$ at 300 GHz, rising to $20\,dB\,cm^{-1}$ at 600 GHz, and over $60\,dB\,cm^{-1}$ at 1000 GHz. In some cases the use of a material which scatters some of the incident signal field may lead to problems. Signal power may be scattered in unwanted directions and, similarly, power coming from directions well away from the intended input beam may be scattered into the output. However, when scattering is acceptable, loaded materials of this general type make very effective low-pass filters.

Quite recently it was noticed that the dielectric properties of Flurogold are polarization sensitive. This appears to be due to a tendency for the suspended pieces of glass fibre to be aligned along a specific axis. For a typical sample the refractive index is 1.605 for the electric field component perpendicular to and 1.625 for the electric field component parallel to the grain axis. The transmission loss for an electric field parallel to the grain axis is around 20% higher than for the perpendicular component. (The values of transmission loss quoted above are average values for randomly polarized radiation.)

In many cases the use of a polarization-sensitive filter would be unacceptable. Fortunately, alternative materials are available. A typical example being Flurosint (a trademark of the Polypenco company). This material consists of a PTFE material loaded with mica granules. Measurements upon samples of this material show no signs of a significant level of polarization sensitivity. A typical sample has a refractive index of around 1.88. The transmission losses in this material are similar to the average values found for Flurogold, being $1 \, \text{dB} \, \text{cm}^{-1}$ at 150 GHz, $6 \, \text{dB} \, \text{cm}^{-1}$ at 300 GHz and $22 \, \text{dB} \, \text{cm}^{-1}$ at 600 GHz. Flurosint also has the advantage of being a strong hard material which is fairly easy to machine using conventional lathe and mill techniques.

Chapter 9

The Martin–Puplett Interferometer

The *Martin–Puplett interferometer* (MPI) is a polarizing version of the *Michelson interferometer*. The incoming signal is split into two equal components which are recombined after being directed along paths of differing lengths. The resulting output signal is therefore modified in a way which depends upon the path difference and the signal wavelength. As each of these systems employs the interference which arises when signals are communicated via two separate beam paths, both are referred to as examples of 'two-beam' interferometers.

Systems of this general type are important for three reasons. Firstly, their performance can be accurately analysed using Gaussian beam mode techniques. Secondly, high-performance MPI systems, able to operate over the whole 75–1000 GHz range, are relatively simple to build and are therefore widely used for millimetre-wave applications. Finally, they are useful as examples of one of the simplest forms of millimetre-wave optical 'circuit'. The MPI hence serves as a useful introduction to the design of millimetre-wave optical instruments.

9.1 The Two-beam (Michelson) Interferometer

In many ways the MPI may be regarded as a development of the Michelson interferometer, designed to overcome some of its drawbacks. It is therefore useful to start a consideration of the MPI by looking at its 'older brother'. Figure 9.1 illustrates the action of a Michelson interferometer system.

The main advantage of a Michelson interferometer is its simplicity. It consists only of a pair of plane mirrors and a semi-reflector employed as a 'beam splitter'. Instruments of this type are widely used at signal frequencies from the ultraviolet to the far-infrared regions of the spectrum. Most commonly, they are used as part of *spectrometers* to measure the power–

The Martin–Puplett Interferometer

frequency spectrum of the input signals. Two-beam interferometers are, however, very versatile and are used for a number of other tasks.

Consider the example of an input signal in the form of a coherent plane wave of wavelength λ which produces an incident electric field

$$E = \sin\left(\frac{2\pi ct}{\lambda}\right) \qquad (9.1)$$

at the beam splitter, c being the velocity of light.

```
        ┊
    ── ╱ ── ▯ ⇿
        ┊
        ▼
```

╱ Beam splitter
⇿ Signal source
∥ Plane mirror
--- Signal beam
▮ Detector

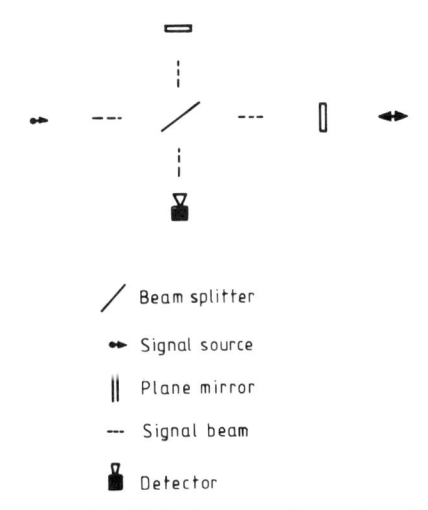

Figure 9.1 Diagram of a Michelson two-beam interferometer.

This will produce a reflected field E_r and a transmitted field E_t:

$$E_r = |r|\sin\left(\frac{2\pi ct}{\lambda} + \phi(r)\right) \qquad (9.2)$$

$$E_t = |t|\sin\left(\frac{2\pi ct}{\lambda} + \phi(t)\right) \qquad (9.3)$$

where $|r|$ and $|t|$ are the magnitudes of the semi-reflector's signal reflectivity and $\phi(r)$ and $\phi(t)$ are the phase shifts introduced upon reflection or transmission.

The reflected and transmitted components of the input signal are directed onto the two plane mirrors which return them to the beam splitter. If the length of the path taken by the reflected component before it arrives back at the beam splitter is Z_1, then the component returns a field

$$E_r(Z_1) = |r|\sin\left(\frac{2\pi ct}{\lambda} + \phi(r) - \frac{2\pi Z_1}{\lambda}\right) \qquad (9.4)$$

to the semi-reflector.

Similarly, if the transmitted component travels a distance Z_2, it produces a returning field

$$E_t(Z_2) = |t| \sin\left(\frac{2\pi ct}{\lambda} + \phi(t) - \frac{2\pi Z_2}{\lambda}\right). \tag{9.5}$$

Part of the $E_r(Z_1)$ field component will be transmitted by the semi-reflector and is directed onto the detector. Similarly, part of $E_t(Z_2)$ is reflected towards the detector. Any power which does not flow towards the detector will, in a loss-free system, be directed back towards the signal source. The total field E_d directed onto the detector and the total field E_s directed back towards the source will be

$$E_d = |t||r| \sin\left(\frac{2\pi ct}{\lambda} + \phi(r) + \phi(t) - \frac{2\pi Z_1}{\lambda}\right)$$

$$+ |r||t| \sin\left(\frac{2\pi ct}{\lambda} + \phi(t) + \phi(r) - \frac{2\pi Z_2}{\lambda}\right) \tag{9.6}$$

$$E_s = |r|^2 \sin\left(\frac{2\pi ct}{\lambda} + 2\phi(r) - \frac{2\pi Z_1}{\lambda}\right) + |t|^2 \sin\left(\frac{2\pi ct}{\lambda} + 2\phi(t) - \frac{2\pi Z_2}{\lambda}\right). \tag{9.7}$$

Using standard trigonometric identities, equation (9.6) can be rewritten as

$$E_d = 2|t||r| \sin\left(\frac{2\pi ct}{\lambda} + \phi(r) + \phi(t) - \frac{\pi(Z_1 + Z_2)}{\lambda}\right) \cos\left(\frac{\pi(Z_1 - Z_2)}{\lambda}\right). \tag{9.8}$$

E_d hence differs in phase from the input field E by an amount $\phi(r) + \phi(t) - \pi(Z_1 + Z_2)/\lambda$ and is attenuated by the factor $2|t||r| \times \cos[\pi(Z_1 - Z_2)/\lambda]$.

In practice, a power detector placed at the output of the instrument will ignore the overall change in phase produced by passage through the system and will measure a mean output power level

$$P = 2|t|^2|r|^2 \cos^2\left(\frac{\pi(Z_1 - Z_2)}{\lambda}\right). \tag{9.9}$$

If the input signal had been directed onto the detector without having to pass through the interferometer system, it would have produced a measured mean power level

$$P_s = \tfrac{1}{2}. \tag{9.10}$$

Hence the efficiency N of signal power coupling through the system is

$$N = 4|t|^2|r|^2 \cos^2\left(\frac{\pi(Z_1 - Z_2)}{\lambda}\right). \tag{9.11}$$

By changing the *path difference* $Z_1 - Z_2$ we may now *modulate* the output power level between the minimum level $N(\min) = 0$, and the maximum level $N(\max) = 4|t|^2|r|^2$. The difference between the maximum and minimum values of N is generally referred to as the *modulation depth*.

From the form of equation (9.11) we can expect that adjacent output maxima and minima will occur when the path difference is changed by λ. Having measured how the modulated output power varies with changing path difference we can therefore obtain a value for the signal wavelength.

Ideally, we would like to obtain a modulation depth M of unity. For a loss-free semi-reflector the values of signal reflectivity and transmissivity will be such that

$$|r|^2 + |t|^2 = 1. \tag{9.12}$$

Hence

$$2|t||r| = 2|t|\sqrt{1-|t|^2}. \tag{9.13}$$

This will reach the maximum possible value if

$$|t| = |r| = \frac{1}{\sqrt{2}} \tag{9.14}$$

when

$$4|t|^2|r|^2 = 1 \tag{9.15}$$

i.e. the modulation depth may (in the absence of any other problems) reach unity if the beam splitter divides the input signal into two components of equal magnitude.

From chapter 8 we know that any real semi-reflector will have a reflectivity and transmissivity which vary in a frequency-dependent manner. This means that the modulation depth obtained will also be frequency dependent when such a semi-reflector is used in a Michelson interferometer.

One of the most common applications of a two-beam interferometer is as a *Fourier transform spectrometer*. Here the variations in detected output power are recorded as the path difference is altered in a controlled manner. The measured variations can then be Fourier transformed to obtain a measure of the signal power–frequency spectrum. In practice the actual calculations are usually performed as a discrete *fast Fourier transform* (FFT). This process, and some of its basic limitations, are outlined in a later section.

For simplicity we can illustrate this process by considering what happens to the detected output power when the path difference is altered at a constant velocity v.

We may think of a complex input signal as being composed of a variety of components, each having an amplitude A and a wavelength λ. Each of these components will produce a sinusoidal variation $P(\lambda)$ in the output detected power of the form

$$P(\lambda) = A^2(\lambda) M(\lambda) \cos^2\left(\frac{\pi v t}{\lambda}\right) \tag{9.16}$$

where

$$M(\lambda) = 4 |t(\lambda)|^2 |r(\lambda)|^2. \tag{9.17}$$

Now, if $M(\lambda) = 1$ at every frequency of interest we may simply determine the detected output fluctuation at the frequency $v/2\lambda$ and use this as a measure of the signal component at the wavelength λ. In practice, however, $M(\lambda)$ will vary with λ. This means that we must independently determine $M(\lambda)$ for every wavelength of interest in order to establish the actual signal spectrum. Worse still, there may be some wavelengths where $M(\lambda) = 0$ and *no* output variations may be observed even if signal power is present.

9.2 Basic Properties of the Martin–Puplett Interferometer

Clearly, unless we only wish to work over a limited signal frequency range, it would be an advantage to find a form of beam splitter whose reflectivity is *not* frequency dependent. Although it is, in practice, impossible to find the ideal of a perfectly loss-free beam splitter whose reflectivity is wholly frequency invariant, there is an alternative which permits us to avoid the problems which arise from the use of conventional beam splitters. The Michelson system may be transformed into an MPI by replacing the semi-reflector with a *polarizer* and the plane mirrors with modified reflectors known as *roof mirrors*.

At millimetre-wave frequencies we can make effective polarizers from plane arrays of closely spaced parallel metal wires. The diameters and spacing of the wires are arranged to be very small compared with the signal wavelength and the wires are many wavelengths in length.

Any field incident upon this array may be regarded as being composed of two orthogonally polarized components, one with its electric vector parallel to the wires, and the other perpendicular to them. Any field parallel to the wires will induce a current along each wire. The resulting current distribution is almost identical with that which the parallel field would set up at the surface of a plane metal mirror. Hence this electric field component is reflected as if the wire array were a metal mirror. The field component perpendicular to the wires is unable to induce a significant current across them and is not reflected. Instead, this field component is transmitted almost as if the wires did not exist. The array of parallel wires therefore acts as a polarizer.

A practical polarizer can be formed using a coil winder of the type commonly used to make transformers and inductors. A flat plate with a pair of frames is mounted on the winder and rotated, winding wire onto the frames. The wires are then glued to the frames and cut free of the plate. For most millimetre-wave applications, tungsten wire of between 5 and 20 μm diameter is used, and the distance between adjacent wires is typically arranged to be about twice the diameter of the wires.

Wire grid polarizers of this type work well over a very wide frequency range. The measured performance of a typical polarizer, wound using 10 μm wire at 25 μm pitch (between wire centres), indicates a cross polar leakage of around -35 dB or less and an absorption–scattering loss of less than -0.01 dB at 250 GHz. This is in fairly good agreement with a simple analysis of an array of wires which can be shown to imply that the grid should have a power reflectivity

$$|r_p|^2 = \left[1 + \left(\frac{2S}{\lambda}\right)^2 \ln\left(\frac{S}{\pi d}\right)^2\right]^{-1} \qquad (9.18)$$

for the electric field component parallel to the wires, and

$$|r_n|^2 = \frac{(\pi^2 d^2)^2}{(2\lambda S)^2 [1 + (\pi^2 d^2)^2/(2\lambda S)^2]} \qquad (9.19)$$

for the component normal to the wires, where S is the spacing between wire centres and d is the wire diameter.

The performance of a wire grid polarizer tends to degrade gently with increasing signal frequency. A typical polarizer with 10 μm wires at 25 μm spacing has a cross polar leakage which falls to around -30 dB at 500 GHz and -25 dB at 1000 GHz. By using 5 μm wires at 10 μm spacing the cross polarization can be reduced to -40 dB at 250 GHz, -35 dB at 500 GHz and -30 dB at 1000 GHz. Even at 1000 GHz the dissipation and scattering losses of these grids is normally well below -0.1 dB. For many purposes, therefore, a well made wire grid may be regarded as a 'perfect' polarizer over the whole millimetre-wave region.

If an incident field, of amplitude A, is plane polarized with its electric vector at an angle θ to the wires, then the transmitted field will have the magnitude $A \sin \theta$ and the reflected field will have the magnitude $A \cos \theta$. By arranging that the input is polarized at 45° to the wires we can obtain transmitted and reflected fields which have the same magnitude $A/\sqrt{2}$, irrespective of the signal frequency. Hence an appropriately angled polarizer will act as a 50:50 beam splitter for a suitable polarized input signal over a wide frequency range.

Unlike the situation which arises with an 'unpolarized' semi-reflector, the reflected and transmitted fields produced when a signal is incident upon an ideal polarizer are 180° out of phase. The transmitted field is simply the field component of the input signal whose electric vector is perpendicular to the wires and hence its phase is unaltered by the presence of the polarizer.

The reflected field component undergoes a phase change of 180° just as it would if the polarizer were replaced by a metal mirror.

For a loss-free 'unpolarized' semi-reflector we may expect that the transmitted and reflected fields undergo phase shifts which ensure that they are 90° out of phase at the plane of the semi-reflector. This can be regarded as a consequence of requiring that the signal power must be conserved and that the fields must be continuous. For a polarizer, these conditions are satisfied by the transmitted and reflected field components being orthogonal, i.e. their fields are orthogonal *in space* instead of being orthogonal in time.

A roof mirror consists of two plane metal reflectors, placed so as to touch along one edge at an angle of 90°. A sketch of a roof mirror is shown in figure 9.2, along with a diagram indicating how it operates. Some workers have referred to this system as a *corner cube*. This usage is, however, both misleading and incorrect. A corner cube consists of *three* plane mirrors placed mutually at right angles so as to form part of a cube. The corner cube arrangement acts as a retro-reflector and its optical properties differ from those of a roof mirror.

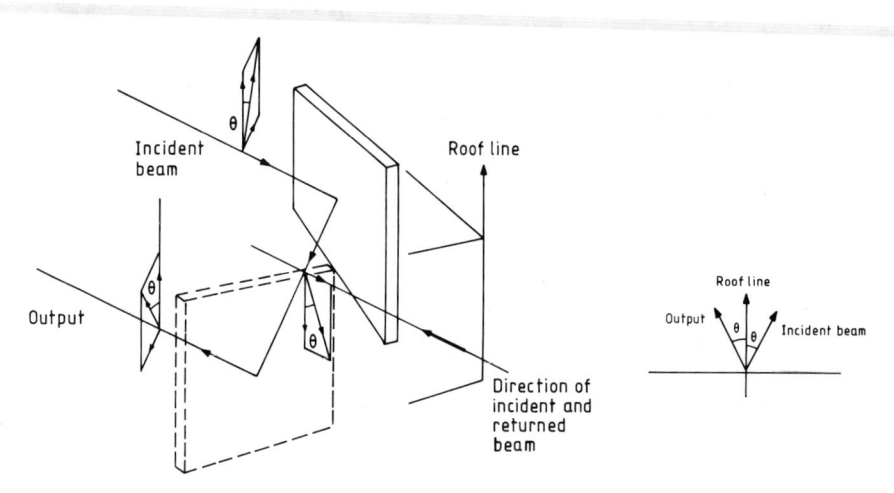

Figure 9.2 Change in polarization produced upon reflection by a pair of plane mirrors located in a 'roof-mirror' arrangement.

The line of intersection of the two mirror planes is called the *roof line*. The usefulness of a roof mirror in polarizing instruments stems from its ability to alter the polarization state of an incident field upon reflection. This process is illustrated in figure 9.2.

A plane polarized signal beam is incident upon one of the mirror surfaces with its electric vector aligned at an angle θ to the roof line. Provided that

The Martin-Puplett Interferometer

the input is incident from a direction normal to the roof line then the output will move away in a direction parallel to the incident beam and its electric vector will be realigned at an angle $-\theta$ to the roof line.

The roof mirror can therefore be used in the place of a normal plane mirror to reflect a signal but has the extra property of being able to 'rotate' the plane of polarization through an angle 2θ. If we now position the roof mirror so that its roof line is at an angle $\theta = 45°$ to the electric field of the input, then the reflected field will have a plane of polarization which has been rotated through $90°$. The incident and reflected fields are now orthogonally polarized.

Figure 9.3 illustrates an arrangement of wire grid polarizers and roof mirrors in the standard form of the MPI. This can be taken as a representation of a suitable arrangement laid out on a horizontal optical table and is convenient as it allows us to refer simply to the orientation of polarizers, fields, etc, with reference to a horizontal plane. In many cases such a system will, in practice, be built as a set of components laid out on such a table. It is important to realize, however, that the diagram may also be taken to represent any equivalent arrangement which may not be horizontal and which may even be 'folded' into more than one plane so as to occupy a convenient physical space more efficiently.

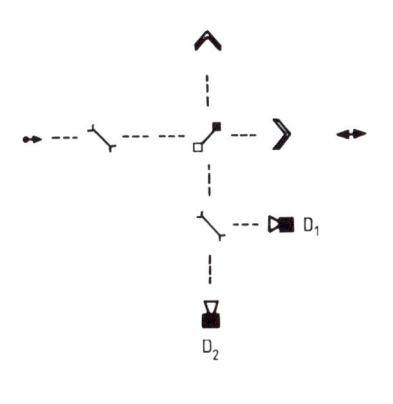

- Signal source
- Polarizer with wires vertical
- Polarizer with wires at 45°
- Fixed roof mirror
- Movable roof mirror
- Detector

Figure 9.3 Diagram of a Martin-Puplett polarizing interferometer.

The illustration can be looked upon as an optical equivalent of a 'circuit diagram' which shows the functional form of an arrangement of basic elements without specifying their physical locations or details. This concept of optical circuit diagrams and their use to understand and design optical systems will be considered in detail in a later chapter.

In the system shown, the beam splitter consists of a polarizer whose wires are aligned to appear at 45° to the horizontal plane *when viewed along the input beam axis*. The input signal is passed through a polarizer which ensures that its field is plane polarized in the horizontal direction. This input polarizer may not be necessary in some cases where the source generates a polarized signal of the desired type.

The 45° polarizer resolves or splits the input signal into two equal, orthogonally polarized components, one being reflected towards a stationary roof mirror and the other being transmitted to a movable roof mirror. Each of these components has its electric field aligned at ±45° to the horizontal plane.

The roof mirrors are positioned with their roof lines in either the horizontal or the vertical plane and aligned normal to the component beam axis. This means that the roof lines are at 45° to the electric fields of the incident components. Hence the reflected components have their field planes rotated through 90°. As a result of this polarization rotation we find that the component which was initially reflected by the beam splitter is now transmitted and the component which was initially transmitted is now reflected. Provided that the polarizer is working essentially perfectly and the roof mirrors are correctly aligned, *all* the signal now moves towards the detectors. *None* of the input power is returned back towards the source.

If we define the unit vectors V and H to represent the vertical and horizontal directions, then an input plane-wave signal E_i of wavelength λ and amplitude A will be of the form

$$E_i = VA \sin\left(\frac{2\pi ct}{\lambda}\right). \tag{9.20}$$

Upon striking the polarizing beam splitter this will generate a reflected field E_r and transmitted field E_t such that

$$E_r = (V + uH)\tfrac{1}{2} A \sin\left(\frac{2\pi ct}{\lambda}\right) \tag{9.21}$$

$$E_t = (V - uH)\tfrac{1}{2} A \sin\left(\frac{2\pi ct}{\lambda}\right) \tag{9.22}$$

where u is either $+1$ or -1 depending upon the orientation chosen for the 45° polarizer. For simplicity, let us take the case when $u = 1$ so that u can be omitted from the following expressions.

If the roof mirrors have their roof lines in the vertical plane, we may

The Martin–Puplett Interferometer

expect that a component polarized in the $V + H$ direction will be reflected with its plane of polarization altered to $V - H$. Similarly, a component incident polarized in the $V - H$ direction will be changed to $V + H$ upon reflection.

The field components returned to the polarizing beam splitter will hence be of the form

$$E'_r = (V - H)\tfrac{1}{2} A \sin\left(\frac{2\pi(ct - Z_1)}{\lambda}\right) \tag{9.23}$$

$$E'_t = (V + H)\tfrac{1}{2} A \sin\left(\frac{2\pi(ct - Z_2)}{\lambda}\right) \tag{9.24}$$

where Z_1 and Z_2 are the distances covered in moving to the roof mirrors and back to the polarizing beam splitter.

The reflected and transmitted components now recombine to produce a total field

$$E_0 = E'_r + E'_t \tag{9.25}$$

which is directed towards the detectors.

Using standard trigonometric identities, E_0 may be written in the form

$$E_0 = VA \sin\left(\frac{\pi(2ct - Z_1 - Z_2)}{\lambda}\right) \cos\left(\frac{\pi(Z_1 - Z_2)}{\lambda}\right)$$

$$+ HA \cos\left(\frac{\pi(2ct - Z_1 - Z_2)}{\lambda}\right) \sin\left(\frac{\pi(Z_1 - Z_2)}{\lambda}\right). \tag{9.26}$$

The MPI differs from the Michelson interferometer in that all the output power is contained in the output E_0. Hence, if we simply detect the mean total output power, we obtain

$$P_0 = \tfrac{1}{2} A^2 \tag{9.27}$$

which is identical with the power of the input beam defined by equation (9.20).

The effect of a non-zero value of the path difference $Z_1 - Z_2$ is to alter the polarization state of the output. In order to observe interferometric modulation, we must therefore measure independently the signal powers contained by the V and H components of E_0. This may be done by using another polarizer to direct the components onto two distinct power detectors which then provide outputs which will indicate the power levels

$$P_V = \tfrac{1}{2} A^2 \cos^2\left(\frac{\pi(Z_1 - Z_2)}{\lambda}\right) \tag{9.28}$$

$$P_H = \tfrac{1}{2} A^2 \sin^2\left(\frac{\pi(Z_1 - Z_2)}{\lambda}\right). \tag{9.29}$$

Both of these powers will vary with $(Z_1 - Z_2)/\lambda$ in a manner similar to that observed when using a Michelson interferometer. However, the polarizing system has two advantages. Firstly, for a well built MPI we do not have the problem of evaluating the effect of variations in beam-splitter performance. Secondly, we have *two* easily available modulated output powers whose sum gives the total input level.

In this particular example a specific alignment was chosen (e.g. vertical roof lines). Similar results would be obtained for different arrangements provided that two general conditions are met. Looking along the signal beam axis we may define an arbitrary pair of orthogonal vectors *a* and *b* perpendicular to the beam axis. The system can then be aligned as follows.

(i) The input polarizer wires, roof lines and output polarizer wires are all parallel to *a* or *b*, when viewed along the beam axis.

(ii) The beam-splitter polarizer wires are parallel to either the *a* + *b* or the *a* − *b* direction, when viewed along the beam axis. The directions *V* and *H* are commonly chosen for these two *principal planes*, but they are not required for the system to perform satisfactorily. Similarly, the choice of which roof mirror should be 'movable' is purely one of convenience.

The system illustrated in figure 9.3 assumes the input signal enters the system by being transmitted through the 'input' polarizer and is hence plane polarized with its electric field in the horizontal direction. It would also be possible to inject a vertically polarized signal by *reflection* from this polarizer. Hence there are a total of *four* distinct ways by which signals may pass in or out of the system. These entrances and exists are generally called *ports*.

The arrangement of optical elements which make up the MPI is symmetric, the distinction between 'input' and 'output' ports being simply one of *use*. A system engineer would therefore find it convenient to regard the MPI as a simple form of *symmetric four-port network*.

In practice we find that a non-zero signal power level will generally enter the system via both input ports. The output modulation observed when the path difference is altered then depends upon the *difference* in the power–frequency spectra of the two signals. If the two inputs are identical, the power modulations that they produce are equal and opposite and no output modulation will be observed.

Although the MPI has most commonly been used as a spectrometer, it does have a number of other applications. For example, the MPI can be employed as an *attenuator*, a *filter*, a *bidirectional cross coupler* (i.e. as a signal *splitter* or *recombiner*), as well as performing the task of a *diplexer* in a heterodyne receiver. In each it is possible to build a system which may be used to provide a high level of performance over the 75–1000 GHz range and which may be tuned easily by changing the selected path difference.

9.3 Gaussian Beam Mode Analysis of the Martin–Puplett Interferometer

Thus far it has been assumed that the signals passing through the interferometer may be simply regarded as plane waves. In order to deal with the diffraction problems which arise in compact systems we can make use of the Gaussian beam mode methods outlined in earlier chapters.

From chapter 3 we may expect that, provided that each optical element has a diameter $d > 3\omega$ (where ω is the beam size at the optical element), then the effects of truncation may for most purposes be taken to be insignificant. For simplicity therefore we can initially assume that the system is built using elements which obey this requirement for any signal of interest.

The action of an MPI can be looked upon as splitting the input beam into two equal components which are directed towards the output via two paths of differing lengths. A given input beam of the form $\psi(z)$ will produce an output beam of the form

$$\Psi(z) = \tfrac{1}{2}[\psi(z + Z_1) + \mathcal{T}\psi(z + Z_2)] \tag{9.30}$$

where Z_1 and Z_2 represent the two differing path lengths and $\mathcal{T}$ is either $+1$ or -1 depending upon our choice of input and output port.

Unless $Z_1 = Z_2$ the two components of $\Psi(z)$ will have different beam sizes and phase-front curvatures. From this it follows that the field pattern of $\Psi(z)$ will not be identical with $\psi(z)$.

The total power-coupling efficiency will be

$$N = \frac{|\langle \Psi(z) | \Psi(z) \rangle|}{|\langle \psi(z) | \psi(z) \rangle|}. \tag{9.31}$$

This represents the efficiency with which power passes from port to port but does not take into account any losses which may arise because the output field may have a modified pattern. When the output from the MPI is to be coupled into an output beam of field $\Phi(z)$, the actual efficiency of coupling will be

$$N' = \frac{|\langle \Psi(z) | \Phi(z) \rangle|^2}{|\langle \psi(z) | \psi(z) \rangle| \, |\langle \Phi(z) | \Phi(z) \rangle|}. \tag{9.32}$$

In most cases we would expect to obtain optimum coupling efficiency when $\Phi(z) \simeq \psi(z)$.

Clearly N' can never be greater than N. In practice we shall generally require both N and N' to be as close as possible to unity. This means that the output components $\psi(z + Z_1)$ and $\psi(z + Z_2)$ should have beam sizes and phase-front curvatures which are fairly similar. This in turn implies that the path difference $Z_1 - Z_2$ should not be too great. Provided that the output pattern $\Phi(z)$ is chosen carefully, N' will then be almost identical with N. For simplicity we can therefore evaluate N and use this as a guide to the effects of beam diffraction upon the performance of the system. In principle we may, of course, seek to arrange that $\Phi(z) = \Psi(z)$ which will ensure that $N = N'$.

For the simplest and most common case where $\psi(z)$ is composed only of the fundamental, circularly symmetric Gaussian mode, it can be shown that

$$N = \tfrac{1}{2}(1 + \mathcal{T} \operatorname{Re}(\beta)) \qquad (9.33)$$

where

$$\beta = \frac{\exp[j2\pi(Z_1 - Z_2)/\lambda]}{1 + j(Z_2 - Z_1)\lambda/2\pi\omega_0^2} \qquad (9.34)$$

for a signal beam of wavelength λ and whose beam waist radius is ω_0. It may be seen that this result depends upon the path difference but *not* upon the location of the beam waist.

Looking at this result we can see that, if $\omega_0^2 \gg (Z_2 - Z_1)\lambda$, then

$$N \simeq \tfrac{1}{2}\{1 + \mathcal{T} \cos[2\pi(Z_1 - Z_2)/\lambda]\}. \qquad (9.35)$$

Now it is a standard trigonometric identity that

$$1 + \cos\theta = 2\cos^2(\tfrac{1}{2}\theta) \qquad 1 - \cos\theta = 2\sin^2(\tfrac{1}{2}\theta).$$

Hence we find that, for a sufficiently large beam waist size, the coupling efficiency N becomes the same as that which would be obtained for a plane parallel signal beam.

More generally, however, the finite value of ω_0 has two immediate consequences, both of which arise from the diffraction of a compact beam.

Firstly, the maximum and minimum values of N will be

$$N(\max) = \tfrac{1}{2}(1 + M) \qquad (9.36)$$

and

$$N(\min) = \tfrac{1}{2}(1 - M) \qquad (9.37)$$

where

$$M \simeq \sqrt{\left[1 + \left(\frac{(Z_2 - Z_1)\lambda}{2\pi\omega_0^2}\right)^2\right]^{-1}}. \qquad (9.38)$$

The parameter M is the *modulation depth*. For a beam whose beam waist size is large enough to be regarded as a plane wave, $M = 1$. For any relatively compact beam, however, M decreases as ω_0 is reduced.

In practice a specific path difference (or range of path differences) will be needed in order for the MPI to be used for a specific purpose. This, along with the signal wavelength, will determine $(Z_2 - Z_1)\lambda$. The value of M—and hence the maximum and minimum values of N—will now be determined by the choice of the beam waist size ω_0.

The second consequence of diffraction is of particular interest when using a two-beam interferometer to measure the signal wavelength or frequency.

The Martin–Puplett Interferometer

An interferometer does not provide a direct measurement of signal frequency. Instead, we must measure how the modulated output power varies as a function of the path difference. This information can then be used to calculate the signal frequency. In order to carry out this calculation we must know how the detected power should alter with path difference for a given signal frequency.

If it is assumed that a signal, of frequency f, passing through the interferometer can be treated as a plane wave, then it follows that the power observed at one of the output ports should vary in proportion to

$$N = \tfrac{1}{2}\{1 + \mathcal{T} \operatorname{Re}[\exp(j\theta)]\} \tag{9.39}$$

where

$$\theta = \frac{2\pi f \Delta}{c} \tag{9.40}$$

$$\Delta = Z_1 - Z_2 \tag{9.41}$$

and c is the velocity of light.

We can define the signal frequency f as

$$f = \frac{c}{2\pi} \frac{\delta \theta}{\delta \Delta}. \tag{9.42}$$

Having measured how N varies with Δ, equation (9.39) can then be used to calculate θ as a function of Δ, and hence a value of f may be obtained.

For a single-mode fundamental Gaussian beam, we know that the term $\exp(j\theta)$ in equation (9.39) must be replaced with

$$\frac{\exp(j\theta)}{1 - j\Delta c/2\pi f \omega_0^2}.$$

This can be rewritten as

$$a \exp(j\theta')$$

where

$$a = \left|\frac{1}{1 - j\alpha\Delta}\right| \tag{9.43}$$

$$\theta' = \theta + \arctan(\alpha\Delta) \tag{9.44}$$

and

$$\alpha = \frac{c}{2\pi f \omega_0^2}. \tag{9.45}$$

The rate of change in θ' with Δ will hence be

$$\frac{\delta \theta'}{\delta \Delta} = \frac{\delta \theta}{\delta \Delta} + \frac{\alpha}{1 + \alpha^2 \Delta^2} \tag{9.46}$$

which means that we have the result

$$f' = f + \delta f \qquad (9.47)$$

where f is the actual signal frequency as defined by equation (9.42) and

$$f' = \frac{c}{2\pi} \frac{\delta\theta'}{\delta\Delta} \qquad (9.48)$$

is the frequency which we would obtain by assuming that the signal may be regarded as a plane wave, i.e. the 'measured' frequency is in error by the amount

$$\delta f = \frac{c}{2\pi} \frac{\alpha}{1 + \alpha^2 \Delta^2}. \qquad (9.49)$$

This error arises because, unlike what would happen if the signal were a plane wave, the output power does not simply vary as a sinusoid with changing path difference. For a plane wave, the rate of phase change as we move parallel to the axis is just $2\pi/\lambda$. For the (m, n)th Gaussian mode, the rate of phase change as we move along the beam axis is $(2\pi/\lambda) + (m + n + 1) \delta\Phi/\delta z$, where Φ is the anomalous phase term. Two-beam interferometry is based upon comparing the relative phases of different places along the beam; hence the output produced is modified by the existence of the anomalous phase term.

A similar analysis may be carried out for beams which consist of a higher-order mode, or a linear combination of modes. The detailed mathematics of this analysis are more involved; however, the results obtained may be understood by considering the general properties of Gaussian modes.

The rate of change in phase along the beam axis depends upon the mode numbers. This means that the frequency 'error' δf made by assuming a plane-wave signal will depend upon the actual mode composition of the beam. Furthermore, since different modes have dissimilar phase rates, their output modulation patterns will not be identical. This will result in some loss of modulation depth as different modes will produce their output maxima and minima at slightly different values of the path difference. For these reasons it is advisable, when using a two-beam interferometer, to arrange that the signal beam consists of just one mode—preferably the fundamental mode. This will maximize the modulation obtained and minimize δf.

9.4 The Limitations of Fast Fourier Transform Spectroscopy

A number of books have been written describing Fourier transform spectroscopy in great detail. Here we shall just consider the basic properties of this method in order to establish the level of performance that we may expect when using a particular system to make spectral measurements. For simplicity in this section we shall assume that the signal may be considered as a plane wave and that the interferometer is well designed and built.

Given the above assumption, an input spectrum of the form

$$S(f) = \int_{f_1}^{f_2} A(f) \cos(2\pi ft) \, df \qquad (9.50)$$

may be shown to produce an output modulated power level which varies as

$$P(\Delta, \mathcal{T}) = \int_{f_1}^{f_2} A^2(f)\tfrac{1}{4}\left[1 + \mathcal{T} \cos\left(\frac{2\pi f\Delta}{c}\right)\right] df \qquad (9.51)$$

where $\mathcal{T} = \pm 1$, depending upon the choice of input and output ports. From equation (9.51) the difference in the modulated output powers at any path difference Δ will be

$$P(\Delta, 1) - P(\Delta, -1) = \int_{f_1}^{f_2} A^2(f)\tfrac{1}{2} \cos\left(\frac{2\pi f\Delta}{c}\right) df. \qquad (9.52)$$

By consulting an appropriate book on the subject of Fourier transforms we may say that for a function which can be defined using an expression of the form

$$G(\Delta) = \int F(f) \cos\left(\frac{2\pi f\Delta}{c}\right) df \qquad (9.53)$$

we may also write that

$$F(f) = \frac{1}{D} \int_{-D}^{D} G(\Delta) \cos\left(\frac{2\pi f\Delta}{c}\right) d\Delta \qquad (9.54)$$

provided that D tends to infinity (i.e. D is very large). Hence we can use the expression

$$A^2(f) = \frac{2}{D} \int_{-D}^{D} [P(\Delta, 1) - P(\Delta, -1)] \cos\left(\frac{2\pi f\Delta}{c}\right) d\Delta \qquad (9.55)$$

to obtain $A^2(f)$ from a measurement of $P(\Delta, 1) - P(\Delta, -1)$ over the range $-D < \Delta < D$. Fourier transformation essentially consists of calculating equation (9.55) for given data and using the results to determine the signal power spectrum, i.e. how $A^2(f)$ varies with f.

In principle, if we know $P(\Delta, \mathcal{T})$ over the whole range $-\infty < \Delta < \infty$, then equation (9.55) provides us with a way of calculating a precise value for $A^2(f)$ at any frequency f. However, in reality, it is impossible to measure $P(\Delta, \mathcal{T})$ over an infinite range of Δ values. In practice we are restricted to altering the path difference over a limited range. This has the effect of reducing the accuracy of the calculated values of $A^2(f)$. The most satisfactory way of viewing this limited accuracy is in terms of a finite spectral *resolution*.

If the range $2D$ were infinite, then each calculated value of $A(f)$ would only depend upon the level of the spectral component at the single frequency f. When $2D$ is finite, each calculated value of $A(f)$ is affected by the signal component levels over a *range* of frequencies centred upon f.

Consider the example of a signal which consists only of a single frequency component $A(f)$ at the frequency f, but for which $P(\Delta, \mathcal{T})$ is only known over the range from $-D$ to $+D$. Using equation (9.55), the calculated values $A'(f')$ that we obtain for the signal magnitude at a frequency f' can be shown to be

$$A'(f') = A(f) \frac{\sin x}{x} \qquad (9.56)$$

where

$$x = \frac{2\pi 2 D(f - f')}{c}. \qquad (9.57)$$

For most purposes we can regard $(\sin x)/x \simeq 0$ when $|x| > \tfrac{1}{2}\pi$. This is equivalent to presuming that $A'(f') \simeq 0$ outside the range $f - c/8D < f' < f + c/8D$.

If the range of path differences D were infinite, we would find that $A'(f') = 0$ for any frequency not equal to f. The result of a finite value of D is to cause the signal at the frequency f to influence the calculated values that we obtain at nearby frequencies. More generally, for some other form of signal spectrum, this is equivalent to saying that the measured power at any frequency will be affected by the signal level at nearby frequencies.

We may regard $c/4D$ as the *resolution* of the measurement as any value $A(f)$ that we obtain will mainly be affected by spectral components within this range, centred upon the frequency f. This means that, for a continuous unpredictable spectrum, we can generally distinguish only components whose signal frequencies differ by an amount comparable with, or greater than, this resolution.

It is important to note, however, that this limitation does *not* mean that we can never measure the frequency of a signal with greater accuracy than $c/4D$. Consider once again the simple case where the signal is known to consist of a single frequency component $A\cos(2\pi ft)$.

Having measured the output power modulation $P(\Delta, \mathcal{T})$ over the finite range $-D < \Delta < D$ of path difference values we can use equation (9.55) to calculate values for $A(f)$. By choosing to carry out this calculation for a series of values which are spread out across the resolution range from $f - c/8D$ to $f + c/8D$, we discover that results obtained vary according to equation (9.56). We may now fit a $(\sin x)/x$ function to these values and locate the frequency at which $A(f)$ is a maximum with an accuracy which

The Martin–Puplett Interferometer

can exceed $c/4D$. In this way the actual signal frequency f may be determined with much greater accuracy than the resolution $c/4D$.

This improvement in the accuracy of the final frequency measurement (and, of course, the correct magnitude of the signal) depends upon being able to perform some extra computations based upon some independent 'knowledge' of the form of the signal, i.e. in this case we expect the signal to consist of a single frequency component. The improvement obtained also relies upon our being confident that any random noise or distortion which has been imposed upon the measurement level $P(\Delta, \mathcal{T})$ is small enough to be ignored.

In principle, this process can be extended to a signal which consists of many, closely spaced frequency components. However, the extra computation required grows rapidly to unmanageable proportions. Furthermore, the effects of unwanted noise or distortions in the measurement process soon limit the extra precision which may be obtained in this way. Hence the process is rarely employed for signals which are not of a fairly simple form.

Any real signal will unavoidably satisfy the following conditions.

(i) The measurement only extends over a finite time or interval.
(ii) The signal is of finite size.
(iii) There will be some random noise superimposed upon the signal.
(iv) There will be an upper limit to the range of signal frequencies present.

Taken together, these conditions mean than any real measurement will contain a finite amount of information. This is why, for a signal of 'unknown' form, the spectral resolution we may obtain is limited.

For the calculations outlined above, it was implicitly assumed that it was possible to measure the power modulation function $P(\Delta, \mathcal{T})$ continuously over the chosen range of path differences. In reality, we measure $P(\Delta, \mathcal{T})$ at a discrete set of locations which are equally spaced some distance d apart. Our measurement then consists of a series of values $P(md, \mathcal{T})$, where m is an integer which identifies the particular value. $P(md, \mathcal{T})$ is generally described as a set of sampled values taken of the continuous function $P(\Delta, \mathcal{T})$.

Knowing that the amount of information measured *must* be finite, irrespective of whether we have measured $P(\Delta, \mathcal{T})$ or $P(md, \mathcal{T})$, we may invoke the *sampling theorem* provided that we arrange to satisfy the two following conditions.

(i) For a signal where λ corresponds to the shortest component wavelength present, the change in path difference d between adjacent samples must be less than $\frac{1}{2}\lambda$.

(ii) The samples must be measured sufficiently accurately that their accuracy is limited by the random noise level present on the signal.

Given that these conditions are satisfied we may make what is referred to as a *complete* record. The amount of information contained by the set of sampled values $P(md, \mathcal{T})$ is now identical with the amount of information that we could have obtained if $P(\Delta, \mathcal{T})$ had been measured continuously. Once a complete set of data samples has been measured, the calculation may now proceed by replacing the integrals shown above with equivalent series. It is important to note that *no* information need be lost owing to the use of a complete series of samples instead of a continuous function.

In practice, modern systems rarely actually calculate the Fourier transform by the basic methods outlined above. Instead a variety of ingenious computational schemes have been devised to reduce the amount of computation required. These computational methods are referred to as *fast Fourier transforms* (FFTs) to distinguish them from the basic method. A typical FFT routine can perform the required calculations in from a tenth to a thousandth of the time required to compute the 'standard' Fourier transform of a signal spectrum. Hence they have replaced the simple Fourier transform for all but a few special applications.

As the numerical techniques involved are varied and often quite complex, we shall avoid discussing how FFT routines operate. Instead, we may restrict our attention to a few basic points concerning their use.

The main extra requirements which are imposed by the use of an FFT method are that the number of input data samples $P(md, \mathcal{T})$ must be 2^n, where n is an integer and the samples must be equally spaced. The output spectrum then consists of the signal magnitudes $A^2(mf_0)$, where $m = 0$ to $\frac{1}{2}n$, and f_0 corresponds to the frequency which would fit just one cycle into the path difference range covered by the set of samples. (It is possible to devise FFT algorithms which operate on data which has not been obtained under these conditions but, in general, the resulting calculation is more involved and takes longer to perform.)

Provided that the FFT is correctly calculated and the input data samples are a complete record, the spectrum obtained should contain all the information which would have been obtained by computing a simple Fourier transform of a continuously measured input, i.e. the FFT method does *not* inherently produce any information loss. Just as in the case with a continuous measurement and a simple Fourier transform it remains possible to use the output of an FFT routine to obtain specific results. It is, for example, possible to use the 'fitting' method to locate the peak of a $(\sin x)/x$ function and hence to identify the frequency of a monochromatic signal with arbitrary accuracy.

With care, Fourier transform and FFT spectra can obtain results whose accuracy is limited mainly by the input signal-to-noise ratio and the interval over which the path difference is altered. In most cases the output from the chosen FFT routine is used as the measured signal spectrum with little or no

The Martin–Puplett Interferometer

further computation. Using a modern digital computer this allows signal spectra to be obtained with minimal delay.

Two-beam interferometers also make it possible to measure the signal levels at a whole range of signal frequencies simultaneously. This is in contrast with systems based upon Fabry–Perot or similar filters which only measure the signal level at one or, at most, a few frequencies at any one time.

In principle, all of the signal power passing through an MPI may be coupled onto the two output detectors and hence contributes to the measurement. The ability to work in this way is generally referred to as the *multiplex advantage*.

Chapter 10

The Design of Optical Circuits

10.1 Circuit Symbols and Design

Electronics engineers make considerable use of *circuit diagrams* to record information on complex arrangements of electronic components. These diagrams can then be used to communicate information as well as providing a medium for a designer to evolve new systems on paper. Circuit diagrams are a form of written *language* which has been devised specifically to record information regarding the function of systems which may be far too involved for a simple explanation in ordinary English.

In the earlier sections of this book, attention was concentrated upon understanding the behaviour of specific devices or of fairly simple optical arrangements. This chapter is devoted to outlining methods which can be used to devise more complex *systems* or *instruments*. In order to achieve this we need to establish two fundamental 'tools'. The first of these is a suitable written language, analogous to conventional electronic diagrams, which will enable us to record the essential details of an optical system. The second requirement is a suitable form of *algebra* which will enable us to calculate the actual functions which a given system will perform.

A circuit diagram does not usually specify every physical detail of the circuit that it describes. The shapes, sizes and physical arrangement of a set of components on a printed-circuit board may be very different from the arrangement of their symbols on a diagram which represents the circuit. Similarly, the diagram may omit information on, say, the voltages and currents in particular places. The diagram is a *stylized* representation which is schematic rather than being realistic. Optical circuit diagrams can be employed in the same manner to record the essential properties of a system without having to detail any real sizes, shapes, etc.

The optical beams employed for millimetre-wave systems differ from conventional electronic cables and microwave transmission lines in one

The Design of Optical Circuits

important respect. They may transmit signals in varying polarization states. This permits us to process signals efficiently using systems which specifically act upon the polarization state of the signal. (The MPI is an excellent example of an instrument which exploits polarization to achieve its function.) It is therefore important to ensure that our circuit language and algebra should enable us to deal with signal polarization and its modification by a system.

Figures 9.1 and 9.3 are examples of optical circuit diagrams, used to represent the Michelson interferometer and MPI. These introduced a specific set of symbols for some of the optical elements that we require. A fuller set of the standardized symbols used in this book is given in appendix 2. As was mentioned in chapter 9, we can regard the circuit as being laid out in a horizontal plane, as if placed on an optical table.

The directions of electric field components, polarizer wire alignments, etc, can then be specified in terms of two vectors V and H. The vector V is defined to be vertical, i.e. V is perpendicular to the nominal plane of the circuit. The vector H is defined to point in the horizontal direction perpendicular to both V and the axis of propagation of the signal beam.

For most purposes we require only a small number of circuit element symbols to define the function of a system. These can be summarized as follows.

Four basic types of *fixed polarizer*, with their wires aligned at $0°$, $90°$, $+45°$ and $-45°$ to the V or H directions *when viewed along the beam axis*, are used. A *variable* polarizer which may be rotated so as to change the angle between its wires and the V or H direction is also useful for many applications.

Generally, only one type of *roof mirror* is required, with its roof line parallel to, say, the H direction.

Symbols are also required for a *plane mirror* placed normal to the signal beam axis, an *absorber* (sometimes called a *beam dump* or *load*) and a *semi-reflector* (which may, of course, have a reflectivity of unity).

To indicate the use of a system it is also useful to employ symbols representing the signal sources and detectors. It is, however, a matter of choice whether or not to include symbols which indicate the presence of focusing lenses or mirrors. This is because these elements do not change the *function* of the system, although they will, of course, affect the practical performance.

A range of other symbols (e.g. one representing a piece of material which acts as a 'lossy' transmitter) can be used for specific applications. The reason why so few symbols are usually needed can be explained by considering these systems as *information-processing* arrangements and comparing them with electronic 'logic circuits'.

In digital electronics it can be demonstrated that only *two* different types of logical element are required to build any other more complex logical system. Even the most complex computer system can be analysed in terms of a large assembly of just two different sorts of 'building block'. For digital logic the basic requirement is that the two basic types chosen must be orthogonal, i.e. you cannot make one of them by an assembly made up from just the other type. (Both the basic types must, of course, actually *do* something which can be clearly defined.)

The most commonly chosen basic building blocks in digital logic are the 'NAND' (not-and) gate and the 'OR' gate. These may then be used to make everything else, from simple 1-bit adders to supercomputers. In practice, of course, it is easier to build a computer using pre-fabricated parts, i.e. you would buy a microprocessor chip and some random-access memory chips rather than build them from scratch using NAND and OR gates, but in principle you could stubbornly build the whole assembly using just the two basic sorts of logical element.

In the same way, optical systems of arbitrary complexity may be patiently assembled from just a few basic polarizers, mirrors and roof mirrors, and some particular arrangements prove useful as 'pre-fabricated parts' for a wide range of applications. As a consequence of this underlying simplicity it is possible to identify circuits which are *synonyms* of each other—different arrangements which perform the same function. Digital engineers are also aware that *redundancy* may often arise in complex systems which have been designed on a simple step-by-step basis, i.e. different parts of the circuit may generate the same signal function for different purposes. Once this redundancy is identified, the circuit can be simplified and the redundant elements removed. This process may also be used to *optimize* or *minimize* optical systems in a similar manner.

Figure 10.1 illustrates three examples of simple arrangements which can be used as building blocks to construct more complex systems.

The properties of each of these arrangements can be seen by considering the behaviour of their elements. In the first example (figure 10.1(*a*)) the input signal is transmitted through a polarizer whose wires are at $+45°$. The transmitted signal, plane polarized at $45°$, is directed onto the roof mirror whose roof line is horizontal. This reflects the signal back towards the polarizer, having rotated its polarization through $90°$. The returned beam is now reflected by the polarizer towards a detector. By moving the roof mirror along the beam axis we can alter the path length between the source and the detector. The arrangement hence acts as a controlled variable time delay. At a specific signal frequency or wavelength this can also be regarded as a phase shifter.

Figure 10.1(*b*) illustrates a rotatable polarizer used as an attenuator. By adjusting the angle of the wires the amount of power directed onto the detector can be changed. In the system shown, unwanted power is directed onto an absorbing load. If this absorber is replaced by a second detector, the system can be used as a variable signal splitter, diverting a chosen

The Design of Optical Circuits

fraction of the total input to each detector output. The arrangement can also be used 'in reverse' to combine two signal components and to direct them along the same output path.

Figure 10.1(c) illustrates a circuit which may be used as an optical equivalent of a controlled complex load terminating a transmission line. The input signal passes through the initial polarizer which has its wires aligned parallel to the horizontal H. The rotatable polarizer then splits the signal power. One component is absorbed by a load, and the other is directed onto a plane mirror which returns power through the angled polarizer.

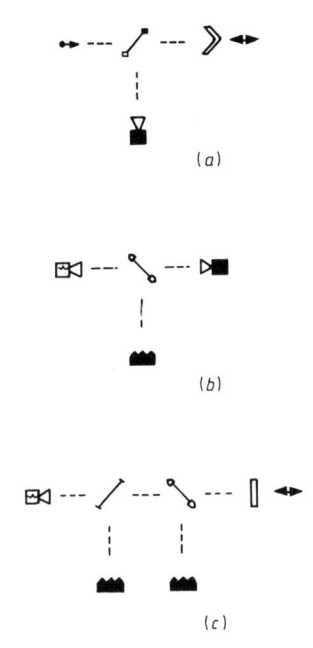

Figure 10.1 Some examples of simple optical circuits. (a) Phase shifter–time delay. (b) Variable attenuator–splitter. (c) Complex terminating impedance.

At first examination it may appear that the fixed horizontal polarizer is redundant as it passes the input signal without loss. However, if it is removed, the polarization of the signal component reflected back onto the source will depend upon the angle of the rotatable polarizer. The fixed polarizer diverts the H component of the reflected signal onto an absorber ensuring that the input and output are both polarized in the same plane.

By rotating the adjustable polarizer we can now change the modulus of the circuit reflectivity between unity and zero. By moving the plane mirror along the beam axis we can alter the phase of the reflected signal. Hence the system may be used to obtain any required value of Γ, the complex signal reflectivity, effectively synthesizing the effect of an arbitrary complex load. In fact, figure 10.1(c) can be regarded as a combination of two similar arrangements, one acting as a phase shifter and the other as an attenuator.

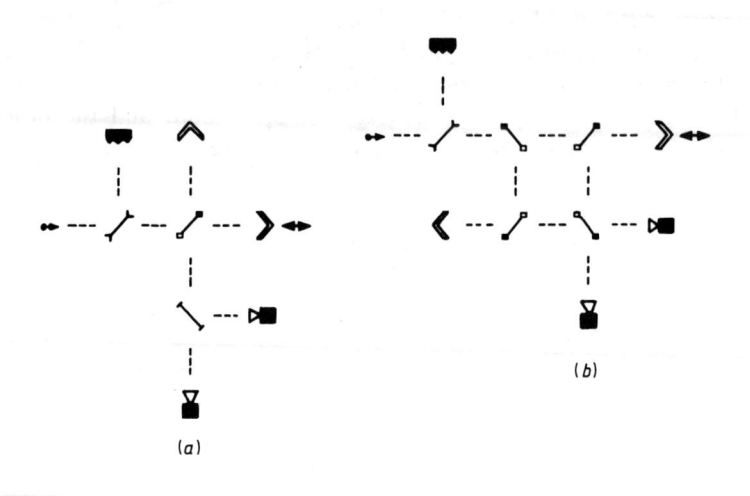

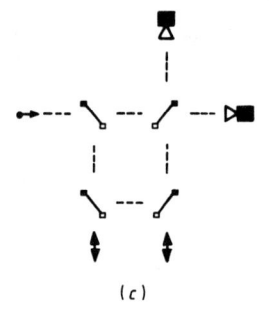

Figure 10.2 Synonyms of the polarizing two-beam interferometer.

Figure 10.2 illustrates three alternative forms of polarizing interferometer. These three arrangements, although different in appearance, are all *synonyms* of the same basic arrangement. In conceptual terms the choice is arbitrary (although there may, of course, be a practical reason for choosing one in preference to the others). Each system splits the input into two orthogonally polarized components of equal magnitudes. The two components are then recombined after being sent along different paths. The output polarization state is hence modulated in a way which depends upon the path difference compared with the signal wavelength.

The examples shown in figure 10.2 represent three different arrangements

The Design of Optical Circuits

which perform the same function. Similarly, we can make a number of minor changes to these circuits (e.g. changing the input polarizer in figure 10.2(a) from vertical to horizontal and the output from horizontal to vertical) without changing their function. By looking for synonyms of basic arrangements we can often simplify the final system and cut down on the number of optical elements required.

10.2 Polarization States and Matrix Algebra

Diagrams of the form illustrated in figures 10.1 and 10.2 are useful 'visual aids' and can be very helpful in developing circuits. They cannot, however, satisfy all our requirements. The diagrams are *qualitative* rather than *quantitative*. To proceed further we require a suitable computational method in order to obtain specific results. To do this we may employ a method based upon the use of *Jones matrices*.

An input signal can be defined in terms of a vector

$$E = E_V V + E_H H \qquad (10.1)$$

which can also be written in the form

$$E = (E_V, E_H). \qquad (10.2)$$

For a coherent signal of frequency f, we can say that, at a given plane,

$$E_V = a \exp(j2\pi ft) \qquad (10.3)$$
$$E_H = b \exp(j2\pi ft + \phi) \qquad (10.4)$$

where a and b represent the magnitudes of the two components and ϕ is the phase difference between them. We can now describe the effect of reflection or transmission by an optical element in terms of a suitable matrix **A** where the output signal field vector O may be obtained from the calculation

$$O = \mathbf{A}E. \qquad (10.5)$$

For clarity, here we shall employ subscripts and use $\mathbf{A_T}$ to indicate transmission through the optical element A, and $\mathbf{A_R}$ to indicate reflection by A. The output produced by, say, reflection from A, followed by transmission through B, and then transmission through C, can be calculated from the expression

$$O = \mathbf{C_T B_T A_R} E. \qquad (10.6)$$

The values of the transmission and reflection matrices of each optical element may be defined from their effect upon an incident polarized signal, represented by a vector E. To define the matrices required, we can start from the simple example of a plane metallic mirror.

The input signal will produce a reflected wave which moves off in a direction which *differs* from the direction of propagation of the input. Since V and H must both be perpendicular to the direction of propagation, we must redefine these vectors for the reflected signal. The convention adopted here is that the vector V shall everywhere be normal to the instrument plane and point 'upwards', i.e. V is not affected by reflection. The vector H is everywhere defined by a 'right-hand law', to have the same orientation with respect to V and the direction of propagation. Hence the direction of H will change upon reflection.

For an ideal metallic surface the reflected and incident fields must be equal and opposite at the surface. Combining this requirement with the redefinition of the direction H upon reflection we can say that the output field O produced by an input signal $E = (E_V, E_H)$ will be $O = (-E_V, E_H)$; hence we can write that the reflection matrix M for a plane mirror will be

$$\begin{pmatrix} -1 & 0 \\ 0 & 1 \end{pmatrix} = M. \tag{10.7}$$

A polarizer whose wires are vertical will only reflect the V component of the incident signal and will transmit the H component as if it were absent. Hence for a polarizer of this type we may define the reflection and transmission matrices

$$\begin{pmatrix} -1 & 0 \\ 0 & 0 \end{pmatrix} = V_R \tag{10.8}$$

$$\begin{pmatrix} 0 & 0 \\ 0 & 1 \end{pmatrix} = V_T. \tag{10.9}$$

Similarly, for a polarizer whose wires are horizontal we can define the reflection and transmission matrices

$$\begin{pmatrix} 0 & 0 \\ 0 & 1 \end{pmatrix} = H_R \tag{10.10}$$

$$\begin{pmatrix} 1 & 0 \\ 0 & 0 \end{pmatrix} = H_T. \tag{10.11}$$

If we define the $+45°$ wire angle so as to appear parallel to $V + H$ when viewed along the beam axis, then we can also show that, for polarizers whose wires are aligned at $+45°$, the reflection and transmission matrices will be

$$\frac{1}{2}\begin{pmatrix} -1 & -1 \\ 1 & 1 \end{pmatrix} = P_R \tag{10.12}$$

$$\frac{1}{2}\begin{pmatrix} 1 & -1 \\ -1 & 1 \end{pmatrix} = P_T \tag{10.13}$$

and for a polarizer whose wires are at $-45°$

$$\tfrac{1}{2}\begin{pmatrix} -1 & 1 \\ -1 & 1 \end{pmatrix} = \mathbf{N_R} \tag{10.14}$$

$$\tfrac{1}{2}\begin{pmatrix} 1 & 1 \\ 1 & 1 \end{pmatrix} = \mathbf{N_T}. \tag{10.15}$$

It can be shown that, for a polarizer whose wires appear to be set at an angle θ to V when viewed along the beam, the reflection and transmission matrices will be

$$\begin{pmatrix} -\cos^2\theta & -\cos\theta\sin\theta \\ \cos\theta\sin\theta & \sin^2\theta \end{pmatrix} = \mathbf{A_R}(\theta) \tag{10.16}$$

$$\begin{pmatrix} \sin^2\theta & -\cos\theta\sin\theta \\ -\cos\theta\sin\theta & \cos^2\theta \end{pmatrix} = \mathbf{A_T}(\theta). \tag{10.17}$$

In general, a polarizer whose wires appear to be at an angle θ to the vertical when viewed along a given beam axis will appear to be aligned at $-\theta$ when viewed from the other side. We must therefore always be careful to define the relevant angle consistently, i.e. θ should always be defined as viewed from the viewpoint of the approaching signal. This dependence upon the signal's direction of propagation has no effect when $\theta = 0$ or $90°$ but must be taken into account for other polarizers.

Another way of looking at this is to consider the effect of rotating the polarizer by $180°$ in the plane of the circuit. This will invert the sign of θ. The symbols used for $+45°$ and $-45°$ polarizers have therefore been chosen such that each may be obtained from the other by such a rotation.

Having established appropriate matrices for the various polarizers and for a plane mirror it is also useful to provide matrices [d] representing the effect of travelling a distance d, and **R** representing the effect of reflection in a roof mirror with its roof line horizontal.

Using arguments similar to those previously employed, it can be shown that

$$\exp\left(-\frac{2\pi\mathrm{j}d}{\lambda}\right)\begin{pmatrix} 1 & 0 \\ 0 & 1 \end{pmatrix} = [\mathbf{d}] \tag{10.18}$$

and

$$\begin{pmatrix} -1 & 0 \\ 0 & -1 \end{pmatrix} = \mathbf{R}. \tag{10.19}$$

These matrices, and similar ones for the other optical elements, can now be used to evaluate the behaviour of various arrangements.

To see how the matrix method is applied we can take the by now familiar example of the MPI shown in figure 10.2(*a*) and consider how it acts upon a source which inputs a signal field

$$E = (E_V, E_H) \tag{10.20}$$

where

$$E_V = A \exp\left[2\pi j\left(ft - \frac{z}{\lambda}\right)\right] \tag{10.21}$$

and

$$E_H = B \exp\left[2\pi j\left(ft - \frac{z}{\lambda} + \Theta\right)\right] \tag{10.22}$$

i.e. a signal of frequency f and wavelength λ whose initial polarization state is set by the values of A, B and Θ.

The signal is initially directed onto a polarizer whose wires are aligned vertically. The signal transmitted into the system will be

$$\mathbf{V_T}E = (0, E_H). \tag{10.23}$$

This plane polarized signal is split into two components by the $+45°$ angled beam splitter. The reflected component is then reflected by the 'fixed' roof mirror before being transmitted by the polarizer to produce an output

$$C_1 = \mathbf{P_T R P_R V_T} E = 0.5(E_H, -E_H). \tag{10.24}$$

The other component is first transmitted, then reflected by the 'movable' roof mirror and finally reflected by the beam splitter to produce an output

$$C_2 = \mathbf{N_R R P_T V_T} E = 0.5(-E_H, -E_H). \tag{10.25}$$

The signal reflected by the movable roof mirror reapproaches the angled polarizer from the opposite direction to the initial input. This means that the polarizer wires now appear to be aligned at $-45°$ and *not* $+45°$—hence the appearance of $\mathbf{N_R}$ in equation (10.25) and not $\mathbf{P_R}$.

The amount of signal directed back towards the source will consist of the two components

$$\mathbf{P_R R P_R V_T} E = (0, 0) \tag{10.26}$$

and

$$\mathbf{N_T R P_T V_T} E = (0, 0). \tag{10.27}$$

Hence, barring some imperfections in the system, no signal power can be reflected back towards the source irrespective of the polarization state of the input signal.

Any 'common' path lengths will produce identical effects upon the phases of C_1 and C_2 and can therefore be ignored unless we need to know the phase of the total output O compared with the input E. If the path lengths

The Design of Optical Circuits

travelled by C_1 and C_2 are the same, we may simply expect O to equal $C_1 + C_2 = (0, E_H)$ and the output signal is vertically polarized. Any difference in the path lengths will modify the polarization state of the output O.

The only parts of the paths which differ are the portions between the angled polarizer and the roof mirrors. If we choose the distance between the angled polarizer and the 'fixed' roof mirror to be d_1 and the corresponding distance to the 'movable' mirror to be d_2, then the output signal components may be written as

$$C_1 = \mathbf{P_T}[d_1]\mathbf{R}[d_1]\mathbf{P_R}\mathbf{V_T}E \tag{10.28}$$

and

$$C_2 = \mathbf{N_R}[d_2]\mathbf{R}[d_2]\mathbf{P_T}\mathbf{V_T}E. \tag{10.29}$$

Under most circumstances, matrix products are *not* commutative; that is, for two random matrices **A** and **B**, $\mathbf{AB} \neq \mathbf{BA}$ and the order of the matrices in an expression cannot simply be swapped around without affecting the result. The distance matrix **[d]** is an exception to this rule because of its *diagonal scalar* form. The above expressions can therefore be rewritten as

$$C_1 = [2\mathbf{d}_1]\mathbf{P_T R P_R V_T}E$$
$$= 0.5 \exp\left(-\frac{2\pi j 2 d_1}{\lambda}\right)(E_H, -E_H) \tag{10.30}$$

$$C_2 = [2\mathbf{d}_2]\mathbf{N_R R P_T V_T}E$$
$$= 0.5 \exp\left(-\frac{2\pi j 2 d_2}{\lambda}\right)(-E_H, -E_H). \tag{10.31}$$

The output electric field levels reflected and transmitted by the output horizontal polarizer will now be

$$O_V = \mathbf{H_T}(C_1 + C_2)$$
$$= j \sin\left(\frac{2\pi(d_1 - d_2)}{\lambda}\right) \exp\left(-\frac{2\pi j(d_1 + d_2)}{\lambda}\right) E_H \tag{10.32}$$

and

$$O_H = \mathbf{H_R}(C_1 + C_2)$$
$$= \cos\left(\frac{2\pi(d_1 - d_2)}{\lambda}\right) \exp\left(-\frac{2\pi j(d_1 + d_2)}{\lambda}\right) E_H. \tag{10.33}$$

Hence the mean power levels falling upon the two detectors will be

$$P_V = \sin^2\left(\frac{\pi \Delta}{\lambda}\right) P_{in} \tag{10.34}$$

and

$$P_H = \cos^2\left(\frac{\pi\Delta}{\lambda}\right) P_{in} \qquad (10.35)$$

where P_{in} is the mean input power level of the signal component E_H, and Δ is the path difference $2(d_1 - d_2)$.

10.3 An Optical Impedance Measurement Circuit

The behaviour of the Martin–Puplett circuit has, of course, already been examined in a previous chapter. The result obtained above can be seen to agree with that found previously using a descriptive analysis. In order to help to appreciate the value of the combination of optical circuit diagrams and matrix methods let us now consider a less familiar and more complex arrangement—that of an *impedance measurement system*.

Consider a situation where we wish to measure the millimetre-wave impedance of a mixer diode. This will enable us to modify the diode or its mounting circuit in order to ensure that signal power is coupled into the mixer with minimal loss. For illustration we can consider the simple arrangement already mentioned in chapter 7 of a diode placed at the end of a 'whisker' or narrow post which shunts a rectangular waveguide. A back-short may be placed behind the post or diode to optimize power coupling.

Using the methods described in chapter 5 we can say that a diode of impedance Z_d in series with a post of inductance L placed as a shunt across a waveguide of characteristic impedance Z_0 will behave as a terminating impedance

$$Z = \frac{(Z_d + 2\pi jL)jZ_0 \tan(\beta d)}{Z_d + 2\pi jL + jZ_0 \tan(\beta d)} \qquad (10.36)$$

when combined with a back-short located a distance d beyond the diode or post. This results in a signal reflectivity

$$\Gamma = \frac{Z - Z_0}{Z + Z_0}. \qquad (10.37)$$

Hence we can calculate the device impedance Z_d from a measurement of the signal reflectivity Γ.

In principle this measurement process is straightforward, but in practice some problems arise which make accurate measurements difficult to obtain. These problems may be divided into two general classes: firstly, the diode and its mounting structure may be different from or more complex than the model that we have assumed; secondly, any unknown or variable properties of the measurement system will have an uncontrolled effect upon the result.

Some of the problems which arise with unknown devices and mounts may be eased by making a series of measurements as a function of some controlled variable. For example, we may measure how Γ varies with back-short position or the DC bias level applied to the diode. Here, however, we are mainly interested in this measurement as an illustration of the use of optical circuit techniques and we shall ignore the first class of problem to concentrate on devising a suitable measurement system.

At microwave frequencies, measurements of reflectivities (and hence impedances) are routinely carried out using systems based upon conventional rectangular metallic waveguides. A metallic waveguide is not the ideal choice of transmission line for this purpose. The guide is *dispersive* (i.e. the effective guide wavelength is not simply proportional to the reciprocal of the signal frequency) and lossy. Its characteristic impedance also varies with signal frequency.

Provided that the system losses, phase delays, etc, are fixed then they can—in principle, at least—be 'calibrated out' of the final measurement. Clearly, however, it would be preferable to employ a system which had low levels of signal loss, was not dispersive and maintained a well defined characteristic impedance. The problems encountered when using metallic waveguides become more severe as the signal frequency rises. Furthermore, the dispersion and losses may vary with time as, for example, the temperature of the system alters. Variations also occur owing to oxidation and mechanical problems at the flanges which connect different waveguide components.

At frequencies around 100 GHz and above, free-space systems can provide comparatively low levels of signal loss and negligible dispersion. Systems built on a firm 'baseplate' also tend to expand and contract as a uniform system. Provided that the measurement technique exploits interferometric methods, temperature changes produce 'common mode' effects which should not alter the measured results. Hence it is possible to build optical circuits which provide a level of measurement accuracy at millimetre frequencies which would be virtually impossible using a conventional waveguide.

Figure 10.3 illustrates a circuit arrangement which may be used to measure the complex reflection Γ produced by a *one-port device* (i.e. a device which has only one input–output port) such as a mixer diode mounted in a guide and tuned with a back-short.

It is probable that, in practice, some extra measurements would need to be made of, say, the reflection properties of the mount (and the antenna used to couple it into the free-space beams) with the mixer diode absent. Here, however, we shall simply assume that a measurement of Γ may be used without difficulty to obtain the device impedance.

Signal power is injected into the circuit from the coherent, vertically polarized source shown on the left-hand side. d represents the mixer whose impedance we wish to measure. n, 1 and 2 are power detectors. Power may pass through the system and reach the detector n by two routes. The first route is via the movable plane mirror and will return a signal

$$C_r = [d_r] H_R N_T V_T A_T(-\theta) M A_T(\theta) V_T P_T H_T E \qquad (10.38)$$

onto the detector n. Here E is the input signal:

$$E = (E_V, 0) = \exp\left[2\pi j\left(ft - \frac{z}{\lambda}\right)\right] V. \qquad (10.39)$$

d_r is the total length of this path and θ is the angle of the adjustable polarizer.

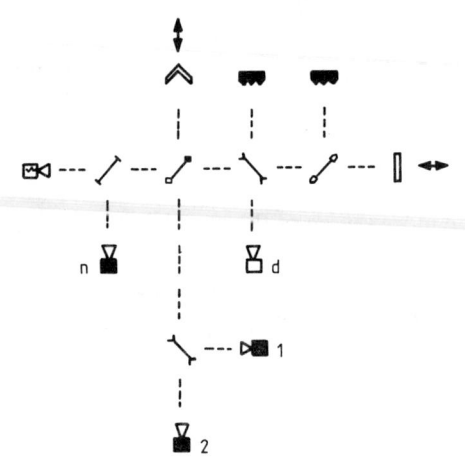

Figure 10.3 Impedance measurement bridge circuit.

The second route will return a signal

$$C_d = [d_x] H_R N_T V_R X V_R P_T H_T E \qquad (10.40)$$

where d_x is the length of this path and $\mathbf{X}$ is a matrix representing the reflectivity of the mixer.

As the signal directed onto the mixer must be plane polarized in the V direction, we may align the mixer and its antenna to couple into this polarization and write that

$$\begin{pmatrix} \Gamma & 0 \\ 0 & 0 \end{pmatrix} = \mathbf{X} \qquad (10.41)$$

where Γ is the complex reflectivity produced by the mixer.

The signal level $C_d + C_r$ falling upon the detector will be zero if we

The Design of Optical Circuits

arrange

$$\mathbf{H_R N_T}([d_r]\mathbf{V_T A_T}(-\theta)\mathbf{MA_T}(\theta)\mathbf{V_T} + [d_x]\mathbf{V_R X V_R})E' = (0,0) \quad (10.42)$$

where

$$E' = \mathbf{P_T H_T}E = (\tfrac{1}{2}E_V, -\tfrac{1}{2}E_V). \quad (10.43)$$

Now

$$[d_r]\mathbf{V_T A_T}(-\theta)\mathbf{MA_T}(\theta)\mathbf{V_T}E' = \exp\left(-\frac{2\pi j d_r}{\lambda}\right)(0, -\tfrac{1}{2}E_V \cos^2\theta) \quad (10.44)$$

and

$$[d_x]\mathbf{V_R X V_R}E' = \exp\left(-\frac{2\pi j d_x}{\lambda}\right)(\tfrac{1}{2}\Gamma E_V, 0) \quad (10.45)$$

and so

$$C_d + C_r = \left[\Gamma \exp\left(-\frac{2\pi j d_x}{\lambda}\right) - \exp\left(-\frac{2\pi j d_r}{\lambda}\right)\cos^2\theta\right]\tfrac{1}{4}E_V. \quad (10.46)$$

The power directed onto the detector will therefore be zero provided that

$$1 = \exp\left(-\frac{2\pi j \Delta}{\lambda}\right)\cos^2\theta \quad (10.47)$$

where Δ is the path difference $d_r - d_x$. We may now calculate Γ from the values of θ and Δ which cause no power to reach the detector n.

The way in which this measurement is made is an example of an important general technique, that of *null* measurement. The instrument compares the effect that we wish to measure with a *reference* which can be adjusted in a well controlled manner. The example described above employs an interferometric approach to act as a *null reflectometer* in order to measure both the magnitude and the phase of the reflectivity Γ.

This approach has many advantages over simpler methods which would relate the property of interest (i.e. the reflectivity) to the measured magnitude of a non-zero power level. For example, the values of θ and Δ which produce a null output do not depend upon the power level generated by the signal source or the sensitivity of the null detector. Hence any fluctuations in the input signal power or uncertainty in the detector's responsivity do not affect the measured result—provided, of course, that the source *is* producing power at a level the detector would be able to sense if θ or Δ is altered away from the 'null setting'. Similarly, any imperfections in the optical and electrical systems which link the source and detector into the instrument will have a 'common mode' effect upon both of the components which are to be compared and hence do not affect the result.

The null reflectometer may be seen as having much in common with a range of other techniques from phase-sensitive detection to the *Wheatstone bridge* used to measure DC resistances.

A particular advantage of the null reflectometer is that, when nulled, no power is reflected back onto the signal source. This may be useful when, as often arises, the source output may be influenced by signal reflections. The nulling arrangement acts as an *isolator*, preventing any reflected power from returning to the source and modifying its output. As a result, circuits of this type can be useful in a wide range of applications.

Having adjusted θ and Δ to obtain zero output from the null detector we may also wish to measure the input or reflected signal levels or the signal frequency. This is valuable when, for example, we wish to determine how the device reflectivity might vary with signal frequency or power. It is also useful to prove that our null result is not caused by a sudden failure of the source or the null detector!

Clearly, the reflected power cannot simply vanish when the null detector observes a zero signal level. Instead we have arranged that all the reflected power is diverted away from the input path by the 45° polarizer and sent towards the other detectors 1 and 2. This is because we have arranged for the circuit to satisfy two conditions. Firstly, the two reflected signals are of equal magnitude. Secondly, they are 180° out of phase.

The total reflected signal returning to the 45° polarizer hence differs in two ways from the component it initially transmitted. Firstly, both its V and its H components are reduced in magnitude by $|\Gamma|$. Secondly, the returned signal is orthogonal to that initially transmitted just as if it had been reflected by a 'lossy' roof mirror.

Signal power may pass to the output detectors 1 and 2 via three distinct paths: via the movable roof mirror, via the device that we are measuring or via the reference plane mirror–rotatable polarizer. The total output directed towards 1 and 2 will hence be a sum of the three components

$$O_r = \mathbf{N_R} \exp\left(-\frac{2\pi j d_r}{\lambda}\right)(0, -\tfrac{1}{2}E_V \cos^2 \theta) \qquad (10.48)$$

via the plane mirror,

$$O_d = \mathbf{N_R} \exp\left(-\frac{2\pi j d_x}{\lambda}\right)(\tfrac{1}{2}\Gamma E_V, 0) \qquad (10.49)$$

via the device, and

$$O_s = [\mathbf{Z}]\mathbf{P_T R P_R}E \qquad (10.50)$$

via the roof mirror, where Z represents the relative length of this path.

When θ and Δ are chosen to satisfy equation (10.47) and produce a zero level at the null detector, we can combine equations (10.48) and (10.49) to obtain

The Design of Optical Circuits

$$O_r + O_d = \exp\left(-\frac{2\pi j d_x}{\lambda}\right)(-\tfrac{1}{2}\Gamma E_V, -\tfrac{1}{2}\Gamma E_V). \tag{10.51}$$

Evaluating equation (10.50) we can also write that

$$O_s = \exp\left(-\frac{2\pi j Z}{\lambda}\right)(\tfrac{1}{2} E_V, -\tfrac{1}{2} E_V). \tag{10.52}$$

The final vertical polarizer will resolve the total output signal $O_s + O_r + O_d$ into two components which will produce mean power levels

$$P_V = \left|\exp\left(-\frac{2\pi j D}{\lambda}\right) - \Gamma\right|^2 \tfrac{1}{4} P_{in} \tag{10.53}$$

and

$$P_H = \left|\exp\left(-\frac{2\pi j D}{\lambda}\right) + \Gamma\right|^2 \tfrac{1}{4} P_{in} \tag{10.54}$$

at the output detectors 1 and 2, where D represents the path difference $Z - d_x$ and P_{in} is the mean power level of the input signal E.

To understand this result more clearly, the above expressions may be rewritten in the form

$$P_V = \left[1 + |\Gamma|^2 - 2|\Gamma|\cos\left(\phi - \frac{2\pi D}{\lambda}\right)\right] \tfrac{1}{4} P_{in} \tag{10.55}$$

and

$$P_H = \left[1 + |\Gamma|^2 + 2|\Gamma|\cos\left(\phi - \frac{2\pi D}{\lambda}\right)\right] \tfrac{1}{4} P_{in} \tag{10.56}$$

where $|\Gamma|$ and ϕ represent the modulus and complex phases of Γ, i.e.

$$\Gamma = |\Gamma| \exp(-j\phi). \tag{10.57}$$

The output power levels P_V and P_H will be modulated sinusoidally as the path difference D is altered. The system behaves as a two-beam polarizing interferometer where each output detector will observe a power level which varies from a minimum of

$$P(\min) = (1 + |\Gamma|^2 - 2|\Gamma|)\tfrac{1}{4} P_{in} \tag{10.58}$$

to a maximum

$$P(\max) = (1 + |\Gamma|^2 + 2|\Gamma|)\tfrac{1}{4} P_{in}. \tag{10.59}$$

By measuring how P_V and P_H vary with Z (and hence with D) we can use standard interferometric arguments to calculate the signal frequency. The sum of the detected powers

$$P_V + P_H = (1 + |\Gamma|^2)\tfrac{1}{2} P_{in} \tag{10.60}$$

may be used to confirm that the source *is* on (!) and is radiating a detectable power level. The difference in the detected powers

$$P_V - P_H = |\Gamma| \cos\left(\phi - \frac{2\pi D}{\lambda}\right) \tfrac{1}{2} P_{in} \qquad (10.61)$$

provides another means of measuring the magnitude and phase of the reflectivity Γ as the magnitude of the modulation of $P_V - P_H$ is proportional to $|\Gamma|$, and the values of Z which produce maxima and minima will be shifted by an amount dependent upon ϕ. The overall measurement system is therefore able to confirm the measurements made by the nulling technique by carrying out interferometry between the nulled reflection and a component of the input signal.

Having examined the properties of just two arrangements, the MPI and a one-port impedance circuit, it should be clear that a wide range of complex circuits may be designed and analysed using the approaches outlined. As in the case in another information-processing area—that of digital logic electronics—these circuits can be assembled from just a few basic elements to perform almost any function a designer wishes to evolve.

10.4 Physical Size and Circuit Bandwidth

The methods described above enable us to design the functional form of a circuit but do not provide any information regarding the practical performance of a real system. Clearly we shall usually wish to build circuits which are as compact as possible without degrading their performance below a required level. Using Gaussian beam mode techniques we can determine the diffraction effects which arise in a real system. Hence it is possible to design a system which can achieve a specific task.

In many cases, real millimetre-wave and far-infrared optical systems have been built using two basic approaches. The first of these makes use of an 'optical table'. Optical elements are placed on a flat table top and may be bolted or pinned in specific locations. The second approach employs 'cubes' as building blocks. The optical table method is particularly useful when we need to make frequent alterations to the circuit arrangement. Cube systems tend to have greater inherent rigidity and are enclosed, making them more suitable for prolonged use in potentially hostile environments.

In recent years a hybrid system, based upon *half-cubes*, has come into increasing use. A half-cube is formed from a hollow cube, cut diagonally into equal halves along a plane containing four cube corners. Optical elements can then be mounted on openings on the two square faces or on the diagonal cut face. The half-cube can then be pinned or bolted to an optical table. Half-cube systems provide a flexible way of being able to

The Design of Optical Circuits

swiftly assemble rigid circuits which contain elements aligned in the required orientations (see photographs 1 and 2).

We may envisage a beam coupled along a circuit path as being passed through a box which has been made from an integer number of open-sided cubes placed end to end. It does not matter whether the beam is reflected by an element placed across the diagonal of a cube or is transmitted to the face opposite. In either case the distance traversed within the cube will be equal to its size L. A beam which passes through n cubes will travel a path length of nL.

Photograph 1 An individual 'half-cube' standing on a baseplate. Optical elements such as lenses, polarizers, etc, may be attached to the faces of a half-cube. An optical circuit can then be assembled by mounting an appropriate arrangement of half-cubes on a flat baseplate. The half-cube shown has a blazed HDPE lens attached to one of its small faces and a wire grid polarizer mounted on its angled face. Dowels on the lower surface of the half-cube are used to locate its position by fitting into an array of holes in the baseplate.

The entrance and exit apertures at each end of the box must have a physical diameter which cannot exceed L. Hence, to avoid truncation problems, the beam size ω at the box ends should ideally be less than or equal to $\frac{1}{3}L$.

Consider now the situation where a lens is placed at the entrance to the box and launches a beam whose parameters are ω and R at the lens. The finite size of the lens will mean that some truncation loss will normally arise at the lens. To avoid any significant further losses due to other apertures within the box (or at the exit), it is desirable to arrange that the beam size at any plane within the box should be less than or equal to ω. In general, however, the performance of most circuits will deteriorate if the beam waist size ω_0 of the beam is reduced. Hence it is also desirable to arrange that ω_0 should be as large as proves convenient.

Photograph 2 A typical millimetre-wave optical circuit assembled using half-cubes. The circuit shown combines a polarizing interferometer with a 'three-mirror' resonator cavity and is intended for noise measurements upon a coherent Gunn oscillator. HDPE lenses, wire grid polarizers, roof mirrors, curved mirrors and feed horns are all shown.

The largest beam waist size consistent with these requirements may be obtained by choosing an input phase-front curvature which causes the waist to be located midway between the input and exit ends of the box. Under these conditions the beam waist size will be

$$\omega_0 = \frac{\omega}{\sqrt{1 + \hat{z}^2}} \qquad (10.62)$$

where

$$\hat{z} = \frac{\lambda z}{\pi \omega_0^2} = \frac{\pi \omega^2}{\lambda R}. \qquad (10.63)$$

The Design of Optical Circuits

Here we may set $z = \frac{1}{2}nL$ and define $\omega = L/u$, where $u \geqslant 3$ to ensure negligible levels of truncation loss.

From the symmetry of such an arrangement it follows that the beam sizes at the input and exit apertures will be identical. Rearranging equations (10.62) and (10.63) we can show that the required beam waist size will be such that

$$\omega_0^2 = \tfrac{1}{2}\omega^2 \pm \sqrt{\tfrac{1}{4}\omega^4 - \left(\frac{\lambda z}{\pi}\right)^2}. \tag{10.64}$$

In order for ω_0 to be real we must therefore ensure that $\omega^2 \geqslant 2\lambda z/\pi$, i.e. the chosen cube side size L must be such that

$$L \geqslant \frac{u^2 n \lambda}{\pi}. \tag{10.65}$$

If this inequality is not met, it becomes impossible to form a beam waist midway between the input and exit faces as the required value of z is too large. The limiting value of z is often referred to as the *maximum throw* of the input beam.

The maximum possible value of λ which produces a real beam waist size for a given choice of n, L and u is often referred to as the *cut-off wavelength* λ_C of the system. From the above we can define this wavelength to be

$$\lambda_C = \frac{L\pi}{u^2 n}. \tag{10.66}$$

This is an analogy to the cut-off phenomenon which arises in a number of forms of waveguide, preventing transmission of signal frequencies below a given value. The analogy is not exact, however, as significant levels of signal power may still be transmitted at signal wavelengths which are regarded as being 'below' cut-off, i.e. when $\lambda > \lambda_C$. These signals will, however, have a beam size at the exit aperture which is larger than at the entrance. Hence the level of truncation loss for a cut-off signal will be largely determined by the exit aperture.

Provided that $L > u^2 n \lambda / \pi$ we may see from equation (10.64) that there will be *two* possible solutions for ω_0. Combining the above expressions these may be written as

$$\omega_p^2 = \frac{L^2}{2u^2}\left[1 + \sqrt{1 - \left(\frac{u^2 n \lambda}{L \pi}\right)^2}\right] \tag{10.67}$$

and

$$\omega_f^2 = \frac{L^2}{2u^2}\left[1 - \sqrt{1 - \left(\frac{u^2 n \lambda}{L \pi}\right)^2}\right]. \tag{10.68}$$

Now in the limiting case where $L \gg u^2 n\lambda/\pi$ these waist sizes become

$$\omega_p^2 \simeq \left(\frac{L}{u}\right)^2 - \left(\frac{un\lambda}{2\pi}\right)^2 \tag{10.69}$$

$$\omega_f^2 \simeq \left(\frac{un\lambda}{2\pi}\right)^2 \tag{10.70}$$

which implies that $\omega_p \simeq \omega \gg \omega_f$ when L is large.

These two values of beam waist size may be distinguished by their behaviour in the limit where L approaches infinity. ω_p is often referred to as the *parallel* beam waist size as it is the choice that we would make when seeking to obtain the largest possible beam waist size in a given situation. Under these conditions the beam may be regarded as being quasi-parallel and approaches a parallel beam as L increases. ω_f is often referred to as the *focused* beam waist as it is the choice we would make when seeking to converge beam power into the smallest possible waist area.

In most cases we choose the largest convenient beam waist size in order to minimize those performance degradations which depend upon beam waist size. Hence a lens or mirror system placed at the input face aperture will generally be designed to position a parallel beam waist ω_p midway between the input and output apertures. Another focusing element can then be placed, if required, at the exit aperture to direct the beam through another set of cubes or to focus the beam into the antenna–feed system of a device.

The beam diffraction which arises between input and output apertures is inevitably dependent upon the signal frequency. In general this will mean that, for a given system, the beam waist size and location will vary with frequency. Hence the signal-coupling efficiency between input and output apertures will normally vary with frequency.

As an illustration of this we can consider the simple case where the beam size ω and phase-front radius R at the input face are frequency invariant. We may arrange that the chosen values of ω and R produced a parallel beam waist midway between the input and output apertures at a particular design wavelength λ_D. By symmetry, at this wavelength the beam at the exit aperture will have a size $\omega' = \omega$ and phase-front radius $R' = -R$. At any other wavelength the exit beam size and radius will differ from these values.

Signal wavelength-dependent losses can now arise in two ways. Firstly, any alteration in ω' will change the truncation losses at the exit aperture. Secondly, alterations in ω' and R' will cause the output beam to be mismatched to the nominally anticipated values of ω and $-R$.

In principle we may arrange that any antenna system coupled to the exit aperture will alter its own beam parameters with signal frequency in a manner which duplicates any variations in the beam output. If this is done, we can overcome the problem of a frequency-dependent beam mismatch. The losses due to aperture truncation will, however, remain.

The Design of Optical Circuits

Using equations (2.33)–(2.40) which were derived to define the beam coupled between two focusing elements, we may say that the required phase-front curvature must be

$$R = \left(\frac{\lambda_C}{\lambda_D}\right)^2 nL\left[1 \pm \sqrt{1 - \left(\frac{\lambda_D}{\lambda_C}\right)^2}\right]. \tag{10.71}$$

At a signal wavelength λ the waist size produced by a beam whose size is L/u at the input aperture and whose curvature is equal to R will be

$$\omega_0(\lambda) = \frac{L/u}{\sqrt{1 + (\lambda_C/\lambda)^2(nL/R)^2}} \tag{10.72}$$

located a distance

$$z(\lambda) = \frac{R}{1 + (\lambda/\lambda_C)^2(R/nL)^2} \tag{10.73}$$

from the input aperture plane.

In most cases we require a parallel beam waist which implies that we should choose the larger value for R which is provided by equation (10.71).

The size of the beam at the exit aperture will be

$$\omega'(\lambda) = \omega_0(\lambda)\sqrt{1 + \left(\frac{\lambda}{\lambda_C}\right)^2\left(\frac{\beta Z'(\lambda)}{\omega_0^2(\lambda)}\right)^2} \tag{10.74}$$

and its phase-front curvature

$$R'(\lambda) = Z'(\lambda)\left[1 + \left(\frac{\lambda_C}{\lambda}\right)^2\left(\frac{\omega_0^2(\lambda)}{\beta Z'(\lambda)}\right)^2\right] \tag{10.75}$$

where $Z'(\lambda) \equiv nL - z(\lambda)$ and $\beta \equiv (L/u)^2/(nL)$.

Hence for chosen values of λ_D/λ_C, nL and L/u we may calculate any truncation and mismatch losses which arise as a function of signal wavelength or frequency.

Assuming that the signal beam is a fundamental Gaussian mode and that the exit aperture radius is $\frac{1}{2}L$ we can use equation (3.30) to say that the power-coupling efficiency when only truncation losses are significant will be

$$T_0 = 1 - \exp\left(-\frac{2(\frac{1}{2}L)^2}{\omega'^2(\lambda)}\right). \tag{10.76}$$

In cases where losses arise owing to beam mismatch at the exit we may use equation (3.40) to obtain the coupling efficiency

$$N_0 = \frac{4\{1 - 2\exp[-(\frac{1}{2}L)^2\alpha]\cos[(\frac{1}{2}L)^2\beta] + \exp[-2(\frac{1}{2}L)^2\alpha]\}}{(\omega_E\omega'(\lambda))(\alpha+\beta)^2} \tag{10.77}$$

where $\alpha = (1/\omega_E)^2 + (1/\omega')^2$ and $\beta = (\pi/\lambda)(1/R_E - 1/R')$. Here ω_E and R_E are the beam size and phase-front curvature at the exit aperture which would ideally match the required output beam.

Figure 10.4 illustrates how the combined mismatch and truncation losses vary with frequency in a particular case where $L = 80$ mm, $u = 3$ and $n = 6$. Under these conditions the cut-off frequency f_C ($=c/\lambda_C$) is 64.4 GHz. The full curve shows the variations in coupling efficiency when a design frequency f_D ($=c/\lambda_D$) of 75 GHz is chosen. This can be compared with the broken curve which is for a design frequency of 150 GHz.

The reduction in coupling efficiency at low frequencies is mainly due to beam truncation. The high-frequency losses are a result of beam mismatch. By comparing the two lines it can be seen that we should choose λ_D to be significantly less than λ_C if a high coupling efficiency is required over a wide frequency range. Conversely, if we require a high efficiency over only a small wavelength range, we may choose values of λ_D and λ_C which are almost identical.

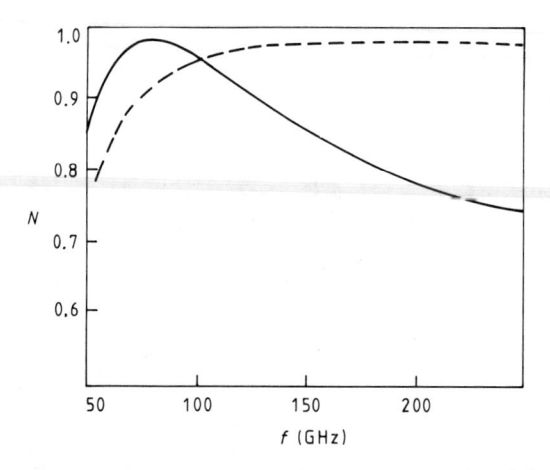

Figure 10.4 Calculated typical losses versus signal frequency in passing through a set of similar cubes of finite size ($L = 80$ mm; $u = 3$; $n = 6$) for two design frequencies: ——, 75 GHz; – – –, 150 GHz.

For a given design wavelength this implies that a broad-band system will require a larger value of λ_C, and hence of L. Hence wide-band systems must, in general, be larger than narrow-band systems.

Ideally, we would only need focusing elements at the beginning and end of an optical path. In practice, however, when n is large, this may require L also to become inconveniently large. To overcome this and to permit a more compact system to be used, we can place extra focusing elements at intervals along the path. When this is done, the system becomes a series of boxes, each having fewer cubes than the total path, reducing the effective value of n which the beam experiences at any place within the system.

The Design of Optical Circuits

This technique of employing a series of—usually, regularly spaced—focusing elements to control the beam is often referred to as a *lens* (or *mirror*) *waveguide*. The system can now be analysed as a series of similar boxes, each n cubes long, in much the same way as above. In this case, however, it is useful to consider the situation where the input (and anticipated output) beam parameters vary in a frequency-dependent manner.

Consider the situation where a large number of similar sets of n cubes are linked together using identical lenses, each having a focal length f.

The refocused input beam into the ith cube set can be taken to have the size $\omega_i(\lambda)$ and curvature $R_i(\lambda)$. This produces an output at the lens which directs the beam into the next cube set whose size and curvature are $\omega_i'(\lambda)$ and $R_i'(\lambda)$. The relationship between these input and output parameters can be defined using arguments identical with those employed above for a single set of cubes.

When considering the behaviour of a single set of n cubes we took as an example the case where the desired beam within the cubes should have its waist plane midway between the input and output ends, i.e. the desired beam within the system was symmetrical. Given identical antennas to couple signals into and out of the system this implies that their lenses would be identical and that the whole beam path is symmetric.

In much the same way, when examining the properties of a series of cube sets, we can assume for simplicity that the input and output antennas are identical. Placing these antennas at equal spacings from the input and output lenses, this implies that optimum coupling through q cube sets will take place when $\omega_1(\lambda) = \omega_q'(\lambda)$ and $R_1(\lambda) = -R_q'(\lambda)$.

Depending upon the values of q, f, etc, it may be shown that more than one choice of $\omega_1(\lambda)$ and f can satisfy this requirement. If, for an integer p, q/p is also an integer, then we can imagine splitting a chain of q cube sets into p smaller cube sets. The above condition can now regarded as being satisfied provided that $\omega_m(\lambda) = \omega_n'(\lambda)$ and $R_m(\lambda) = -R_n'(\lambda)$, where m and n are integer multiples of q/p.

Using this argument we can see that the number of possible solutions which satisfy the required condition will depend upon the number of integer divisors of q. These solutions may be regarded as being *periodic* as the beam profile will repeat itself every q/p cubes along the path. For those cases where q is prime, only two types of solution are possible: those for $p = 1$ and for $p = q$.

In practice we are generally interested in the simplest solution where $p = 1$, i.e. we look for a solution where the beam profile repeats itself in every cube set along the path. Under these circumstances, $R_i(\lambda) = -R_j'(\lambda)$ and $\omega_i(\lambda) = \omega_j(\lambda)$ for any integer values of i and j. Hence $R_i(\lambda) = -R_i'(\lambda)$, and the beam waist plane in each cube set must, by symmetry, be at the middle of the cube set.

The focal length f of each lens will be such that

$$\frac{1}{f} = \frac{1}{R_i'(\lambda)} - \frac{1}{R_{i+1}(\lambda)} \tag{10.78}$$

i.e.

$$f = \tfrac{1}{2} R_i'(\lambda). \tag{10.79}$$

Now, for a given set of lenses, f will have a specific value which is frequency independent. Hence, for a given series of lenses, R_i (and R_i') must also be frequency independent.

As was the case for a single set of cubes we may define the design of a practical system by choosing a cube size L and a beam size $\omega = L/u$ at the lenses for a given design wavelength λ_D. For a cube set which is n cubes long, equation (10.66) can be used to obtain a value for λ_C and the appropriate phase-front radius R_i of curvature can be obtained from equation (10.71). The required lens focal length f can then be obtained from equation (10.79).

Unlike the example considered earlier, both the exit *and* the input beam sizes must now be frequency dependent. For this reason, whilst λ_C is useful in calculating R, it does not represent an actual cut-off wavelength. Combining and rearranging equations (10.62) and (10.63) it can be seen that, when R is frequency independent and the beam waist is located at a distance $\tfrac{1}{2} nL$ from each lens, the beam size $\omega(\lambda)$ at each lens will be such that

$$\omega^4(\lambda) = \frac{\lambda^2 R n L}{2\pi^2 (1 - nL/2R)} \tag{10.80}$$

i.e. $\omega(\lambda)$ is proportional to $\sqrt{\lambda}$.

Hence the required beam size at each lens (and the beam waist size) tends to increase with the signal wavelength. In the absence of any truncation effects caused by finite lens diameter this would mean that—in principle—a coupling efficiency of unity would be possible at arbitrarily long signal wavelengths. In reality, however, any lens must have a finite size and, as λ increases, the truncation losses must become increasingly severe.

In order to minimize ω at any wavelength, we may choose to arrange that $\lambda_D = \lambda_C$. This produces the phase-front curvature $R = nL$ and leads to a beam size

$$\omega(\lambda) = \sqrt{\frac{\lambda n L}{\pi}} \tag{10.81}$$

at each lens.

For this special case, the beam incident upon or departing from a lens has its phase-front curvature centred on the adjacent lens, and the focal length f of each lens is $\tfrac{1}{2} nL$. An arrangement of this type is often referred to as a *confocal* system.

Chapter 11

Oscillators and Signal Sources

On theoretical grounds it makes sense to divide millimetre-wave sources into two general classes: incoherent wide-band sources and coherent oscillators. In practice, however, most sources will be of three identifiable types:

(i) thermal incoherent wide-band sources;
(ii) vacuum state oscillators;
(iii) solid state oscillators.

Vacuum state oscillators such as the *klystron* and the *carcinotron* (or *backward-wave oscillator*) have been, and still are, widely used as coherent millimetre-wave sources. However, they are increasingly being replaced by solid-state oscillators which are smaller, easier to operate and promise better performance with increased reliability. Hence in this chapter we will only deal briefly with vacuum state oscillators before examining solid state oscillators in more detail. Before this, however, it is worth outlining the general properties of thermal sources and their uses.

11.1 Thermal Sources

Ideally a thermal source radiates as a *black body*, i.e. the radiation intensity S at a given signal frequency ν simply obeys the Planck equation

$$S = \frac{2h\nu^3/c^2}{\exp(h\nu/kT) - 1} \qquad (11.1)$$

where S is in units of watts per square metre per hertz per square radian, T is the physical temperature of the source in kelvins, k is Boltzmann's constant, h is Planck's constant and c is the velocity of light.

For signals of frequency $\nu \ll kT/h$, this expression is approximately equal to

$$S = \frac{2kT}{\lambda^2} \tag{11.2}$$

where λ is the signal wavelength.

The *Wien displacement law* states that S will reach a maximum value at a wavelength λ_M such that

$$\lambda_M T = 5.1 \text{ (mm K)}. \tag{11.3}$$

For wavelengths much greater than λ_M, the radiation intensity will be as given by equation (11.2).

Now S represents an amount of power flowing through a unit area in a range of directions contained by a unit of solid angle. Consider the situation where a power detector is connected to a short length of transmission line and signal power is coupled into the other end of the line using an antenna. The antenna has an effective collecting area A_e and receives power directed onto it over a range of directions of total solid angle Ω. If the thermal source is large enough to fill all the directions covered by Ω, then the power received will be

$$P = SA_e\Omega \tag{11.4}$$

per hertz of signal bandwidth. However, $A_e\Omega = \lambda^2$; hence

$$P = 2kT \, \text{Hz}^{-1}. \tag{11.5}$$

In fact the power radiated by the thermal source is incoherent. This means that its polarization state will vary randomly from moment to moment. For most cases of practical interest our detector and its transmission line will respond only to signals of one (usually plane) polarization state. The antenna and detector will hence act as if a polarizer had been placed between the source and the detector. This will, on average, prevent half the total power from reaching the detector.

Hence the total power seen by the detector from a thermal source which fills the antenna's field of view will be

$$P = kTB \tag{11.6}$$

where B is the signal bandwidth to which the detector responds. This result is identical with that which would be obtained if the thermal source and antenna were replaced by a resistor of temperature T whose resistance matches that of the transmission line, i.e. the detector is unable to distinguish between a thermal source radiating into an antenna and a matched resistive load of the same temperature.

Black-body sources have two useful features. Firstly, they emit essentially the same power per unit bandwidth over a very wide frequency range.

Oscillators and Signal Sources

Secondly, they tend to radiate power isotropically and in all polarization states. These properties, combined with a power output which simply varies in proportion to temperature, makes them particularly useful as calibration or reference sources. An example of this form of use has already been mentioned in the chapter on mixer systems where a thermal source is used to define and measure the noise temperature of a heterodyne receiver.

Thermal sources are also widely used in spectroscopy to provide power for measurements of the absorption spectra of various devices or chemicals. An illustration of this use is represented in figure 11.1 where a polarizing interferometer is being used to measure the absorption properties of a sample S.

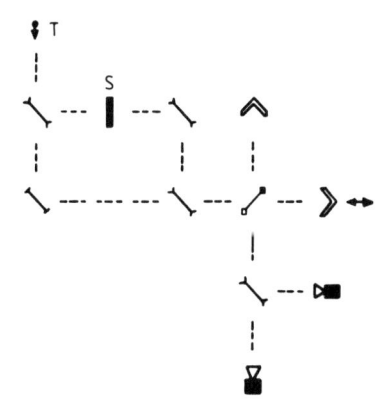

Figure 11.1 Circuit for measuring the effects of signal transmission through a sample S from a source T.

A two-beam interferometric spectrometer measures the *difference* between two signal levels as a function of signal frequency. It is possible in principle to arrange for one signal source to consist of a black body at absolute zero, hence setting one of these signal levels to zero. In all real cases, however, it is necessary to realize that the measurement is comparative and that *neither* power level is zero. The system illustrated in figure 11.1 is an example of how the comparative nature of such a system can be taken into account when making a measurement. The system operates by comparing a signal from a thermal (incoherent unpolarized) source with a signal from the same source which has been modified by passage through the sample S.

Consider thermal power emitted by the source T in a small signal frequency interval dB centred upon the wavelength λ. The source will radiate a horizontally polarized signal power level

$$P_H(\text{in}) = kT \, dB \tag{11.7}$$

which passes through the interferometer to produce the power levels

$$P_H(T) = kT \, dB \, \cos^2\left(\frac{\pi\Delta}{\lambda}\right) \tag{11.8}$$

and

$$P_V(T) = kT \, dB \, \sin^2\left(\frac{\pi\Delta}{\lambda}\right) \tag{11.9}$$

falling upon the two output detectors.

The vertically polarized component of the signal power is directed into the interferometer via the sample S. If the absorptivity of this sample is $\alpha(\lambda)$ at the wavelength λ, then the power which is transmitted will be $(1-\alpha)kT \, dB$.

Now the sample will also *emit* power if its temperature T_S is above absolute zero. The total vertically polarized power level entering the interferometer will therefore be

$$P_V(\text{in}) = (1 - \alpha(\lambda))kT \, dB + \varepsilon(\lambda)kT_S \, dB \tag{11.10}$$

where $\varepsilon(\lambda)$ is the emissivity of the sample at the signal wavelength λ.

In general, the absorptivity and emissivity of a sample are the same, i.e. $\alpha = \varepsilon$ hence this vertically polarized signal will produce power levels

$$P_V(S) = [(1 - \alpha(\lambda))T + \alpha(\lambda)T_S]k \, dB \, \cos^2\left(\frac{\pi\Delta}{\lambda}\right) \tag{11.11}$$

and

$$P_H(S) = [(1 - \alpha(\lambda))T + \alpha(\lambda)T_S]k \, dB \, \sin^2\left(\frac{\pi\Delta}{\lambda}\right) \tag{11.12}$$

falling upon the two detectors.

The difference D in the total power levels falling upon the two output detectors will therefore be

$$D = (P_V(T) + P_V(S)) - (P_H(T) + P_H(S))$$

$$= \left[\cos^2\left(\frac{\pi\Delta}{\lambda}\right) - \sin^2\left(\frac{\pi\Delta}{\lambda}\right)\right]\alpha(\lambda)(T_S - T)$$

$$= \cos\left(\frac{2\pi\Delta}{\lambda}\right)\alpha(\lambda)(T_S - T). \tag{11.13}$$

We can therefore determine $\alpha(\lambda)$ as a function of λ by measuring D as a function of Δ and carrying out a Fourier transform.

It should be noted that the magnitude of the observed modulation in D as Δ is varied is simply proportional to the difference $T_S - T$ in the temperatures. It follows that, if $T_S = T$, then $D = 0$ for every path difference Δ and we will be unable to measure $\alpha(\lambda)$. To maximize the modulation of D with changing Δ, the temperature difference should be as large as is convenient.

Many practical thermal sources take the form of a *cold load*, typically a sheet of microwave absorber cooled by immersion in a liquefied gas, usually liquid nitrogen. Under normal laboratory conditions, liquid-nitrogen cooling provides a wide-band thermal source with a temperature $T = 77$ K. When used to measure the properties of a sample which is at normal room temperature, this provides a frequency difference of around 215 K.

Nitrogen-cooled thermal sources are easy to manufacture and use, liquid nitrogen is widely available and boiling nitrogen provides a reasonably well defined temperature for the source. Hence cold loads of this type have many advantages over *hot loads*, when $T > T_S$. The main disadvantage of cold loads is that the temperature range is clearly restricted to $T \geqslant 0$ K. This may mean that $T - T_S$ is in some cases too small to obtain a reliable signal for measurement.

Hot loads may, in principle, be of any temperature. However, the design, manufacture and use of hot thermal sources is somewhat more difficult than the cold alternative. In particular, it is difficult to arrange for a very hot source whose temperature is uniform and well defined. Traditionally, one of the most common hot loads in millimetre-wave systems has been a mercury lamp. This represents quite a complex source as radiation is emitted from the ionized gas, the glass or quartz tube and the immediate surroundings of the lamp. Each of these regions may have a different temperature and millimetre-wave emissivity.

Mercury lamps also tend to be rather small compared with most free-space beams; hence a suitable antenna is required which can withstand the high temperatures. This contrasts with a nitrogen-cooled load which may simply consist of a sheet of absorber large enough to completely fill the cross section of a free-space beam.

The main advantage of a mercury lamp hot source is that its temperature is quite high and hence it can offer a large value of $T - T_S$ when T_S is around room temperature. As with other types of hot load, however, the accuracy of final measurements may primarily depend upon being able to calibrate the effective temperature of the thermal source.

11.2 Vacuum State Coherent Oscillators

A bewildering variety of vacuum state oscillators have been developed and used as microwave and millimetre-wave oscillators. Fortunately, although the detailed behaviour of these devices varies from one to another, their basic properties have much in common. For the sake of simplicity, therefore, we shall consider just two types which have been commonly used as millimetre-wave sources: the *klystron* and the *backward-wave oscillator*.

Figure 11.2(a) illustrates a klystron. A beam of electrons is directed through two pairs of metal grids (or plates with holes for the beam to pass through). A small oscillatory voltage is applied to the first pair of grids, setting up an alternating electric field between them, parallel to the direction in which the electrons are moving. This alternately accelerates or decelerates the electrons which happen to be travelling between the grids at any particular moment.

The beam leaving this first pair of grids now consists of alternate regions of 'fast' and 'slow' electrons. As these move towards the second pair, there is a tendency for the fast electrons to catch up with the slower ones in front of them. This means that, at a particular distance away from the grids, the initially uniform beam will have been changed into a set of 'bunches' of moving electrons separated by regions which are almost empty.

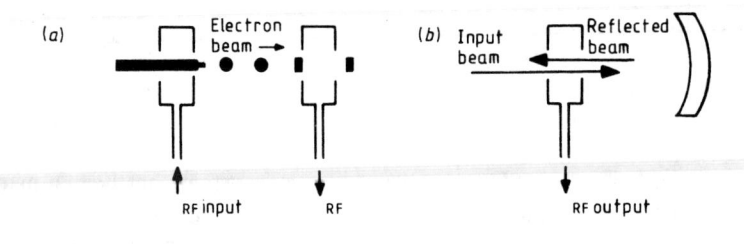

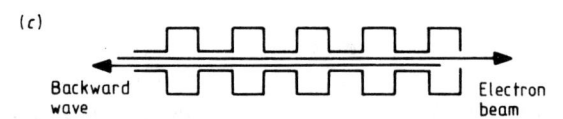

Figure 11.2 Some forms of vacuum device which can be used as electron beam oscillators. (a) Simple klystron. (b) Reflex klystron. (c) Backward-wave oscillator.

A second pair of grids, placed at an appropriate *drift distance* from the first, will receive a stream of electron bunches which arrive at a repetition rate set by the frequency of the oscillatory voltage applied to the first pair. As each bunch passes first one grid and then moves to the other, it changes the potentials on them. The repetitive passage of the bunches hence sets up a field between this second pair of grids which oscillates at the same frequency as the input to the first pair.

Provided that the input beam charge density, velocity, drift distance, etc, are chosen with care, this arrangement acts as an amplifier and we may remove any signal power from the second grid pair which is greater than was injected onto the first pair. This increase in the signal level is achieved by subtracting energy from the electron beam. On average, electrons leaving the second grid pair will have less kinetic energy than when they entered the system.

Now, in principle, any signal amplifier may be used as an oscillator by the application of *feedback*. Almost all real amplifiers employ feedback in order to control the signal amplification; hence the relationship between amplifiers and oscillators is close. (Many engineers regard the following as one of the laws of thermodynamics: the perversity of the universe tends to a maximum. One result of this is that *any amplifier that you build oscillates, but any oscillator that you build turns out to be an amplifier*.)

Consider the example of an amplifier whose signal gain is α which is used with a feedback arrangement which passes a fraction β of the amplified signal back to the input. Both α and β are complex, their complex angles indicating the phase shifts which take place during amplification and feedback. Using the standard arguments of electronics it can be shown that the overall signal gain α' of such a feedback-controlled amplifier will be

$$\alpha' = \frac{\alpha}{1 - \alpha\beta}. \tag{11.14}$$

Clearly, if $\alpha\beta = 1$, then $\alpha' = \infty$, i.e. the signal gain becomes infinite; hence we would require $|\alpha\beta| < 1$ to ensure that the arrangement works as an amplifier whose gain may be defined. As a corollary of this we may expect the arrangement to act as an oscillator if $\alpha\beta \geq 1$.

For any practical system, both α and β will depend upon the signal frequency f. For a system to oscillate at the frequency f we must satisfy the requirement that

$$|\alpha(f)\beta(f)| \geq 1 \qquad \text{angle}(\alpha(f)\beta(f)) = 2\pi n. \tag{11.15}$$

This requirement is sometimes referred to as the *Barkhausen criterion*.

Provided that we ensure that this criterion is only satisfied at a single frequency f, the system may be used as an oscillator to produce coherent output at this frequency.

The klystron amplifier can hence be turned into an oscillator by feeding a small portion of the output signal from the second to the first pair of grids so as to provide the initial varying potential which starts the bunching process. Provided that the klystron amplifier gain is large, there is no need to start the process by injecting a specific signal. Small random variations in the beam electron density will provide an initial noise signal which will be enough to cause oscillation to commence.

The frequency of oscillation will depend upon the time taken for electrons to drift between the pairs of grids and the time taken for the signal to be fed back. This means that the frequency of oscillation is largely set by the physical size of the system, although the oscillator may be tuned over a limited range by changing the beam accelerating voltage, altering the mean velocity of the electrons.

An alternative version of the basic arrangement, called a *reflex klystron* is shown in figure 11.2(*b*). It differs from the simple klystron in having only one grid pair and a negatively charged beam *reflector*. The method of operation is, however, almost identical. The electron beam initially passing between the grids experiences an oscillatory field which starts the process of bunching.

As the beam moves away from the grids, the electrons are repelled by the field from the reflector. The negative charge on this electrode is sufficiently high to prevent the beam from reaching it. Instead, the bunching electrons are reflected back towards the grids. The electron bunches then pass back between the grids, inducing the alternating field which, in turn, starts the process for other electrons moving into the system.

The reflex klystron has two advantages over the simpler system. Firstly, it can be easier to build as only one grid pair is required. Secondly, the frequency of operation may be tuned by altering the reflector potential. As the reflector potential is increased, the electrons are returned to the grids in a shorter time, tending to increase the oscillation frequency. This time can, however, only be altered over a limited range before the electrons returning to the pair of grids cease to be correctly bunched.

In practice, the behaviour of a klystron is more complex than the simple description given here and a number of other factors affect the performance. For example, each grid pair will form a cavity which will resonate at a particular frequency. As a result, the klystron will only be able to oscillate freely at a frequency near this resonant value. In order to widen the accessible frequency range, most millimetre-wave klystrons are therefore manufactured in a manner which allows their size and shape to be altered in order to provide some 'mechanical' frequency tuning.

Figure 11.2(*c*) illustrates the kind of arrangement used in a wide range of vacuum oscillators and amplifiers from *travelling-wave tubes* to *backward-wave oscillators*. Here an electron beam is directed so as to pass through (or nearby) a periodic metallic structure of some kind. The passage of the electron beam can induce currents in the metal surface which, under the right conditions, will produce an electromagnetic wave moving along the structure. Power can then be transferred from the kinetic energy of the electron beam into the electromagnetic wave. This wave may either propagate in the same direction as the electron beam to produce a *forward* wave or propagate in the opposite direction to produce a *backward* wave.

The charges on the electrons will tend, unless corrected, to cause the beam to spread out as it moves along. To help to prevent this a strong magnetic field, aligned parallel to the beam axis, is employed to restrict transverse movement of the electrons. The detailed behaviour of these forms of amplifier–oscillator is generally deduced by considering them as a chain of coupled resonant cavities. For clarity, however, it is more convenient here to consider the structure as an odd form of waveguide.

Oscillators and Signal Sources

In the absence of an electron beam we can imagine an electromagnetic wave propagating along the structure–waveguide. This wave will have a particular field pattern and propagation constant which will depend upon its signal frequency and the details of the structure. In general, we can expect that the waveguide will be dispersive, i.e. the propagation constant and wavelength of the signal in the guide will *not* simply vary proportionally to the signal frequency.

If the waveguide only permits a single mode to propagate at the signal frequency, then the field pattern and propagation constant can be defined uniquely; otherwise the signal may consist of more than one field pattern, each having its own value of propagation constant. For simplicity, here we shall assume that only one mode is possible.

An electron beam passing near the structure will be influenced by any alternating fields propagating along the structure. Under some circumstances the beam will tend to bunch, producing an uneven charge distribution—and hence a varying field—along the beam. This bunching can then alter the field propagating along the structure.

In practice, we can expect four possible types of wave–beam interaction. These can be distinguished according to whether power is transferred from the beam to the wave or from the wave to the beam, and whether the wave is moving forwards or backwards compared with the beam. Interactions which transfer energy to the beam are of interest to those building particle accelerators. Interactions which transfer energy from the beam to the travelling field provide a method of operation which can be exploited to provide amplifiers and oscillators.

Backward-wave oscillators, where power is transferred—as the name implies—to a signal wave moving in the opposite direction to the electron beam, have a much wider electronic tuning range than klystrons. This is because the signal frequency depends strongly upon the mean velocity of the electrons in the beam. The signal frequency obtained from a row of cavities of a given scale size also tends to be smaller than would be produced by a klystron whose resonant cavity was of similar dimensions.

At around 100 GHz a typical reflex klystron can be expected to produce a continuous power level up to about 1 W and will require a beam voltage of 2–3 kV. The electronic tuning range of a reflex klystron is typically of the order of 100 MHz, and the mechanical tuning range will be then between 1 and 10 GHz. The operational life of a millimetre-wave klystron is usually a few hundred hours, although this may be extended by operating the tube at relatively low power levels.

These values may be compared with a typical 100 GHz backward-wave oscillator which can produce around 10 W or more and can be electronically tuned over a bandwidth in excess of 30 GHz. Some backward-wave oscillators have been tested which proved able to produce over 100 W continuous output at 94 GHz. Others have provided continuous power levels of about 1 mW at signal frequencies about 1000 GHz. The operational life of a backward-wave oscillator is generally 10–100 times that of a klystron.

The main disadvantages of the backward-wave oscillator are its relatively large size and the need for a beam voltage of 4–8 kV. It also requires a powerful (and bulky) magnet to maintain a compact electron beam along the length of the internal structure. Whilst excellent for many applications, the backward-wave oscillator is an unsatisfactory choice when the oscillator (and its power supply!) must be small and/or light.

11.3 Solid State Oscillators

The basic properties of solid-state coherent sources can best be described in terms of a *negative resistance* which is coupled to a resonant circuit or cavity.

Consider a closed loop formed by connecting together a capacitance C, inductance L and a resistance R. In order for the potential at any point around the loop to be single valued the current $i(t)$ circulating around the loop at an instant t must satisfy the expression

$$i(t)R + \int_0^t \frac{i(t)}{C} \, dt + L \frac{di(t)}{dt} = 0 \tag{11.16}$$

where the moment $t = 0$ is defined to be a time when the charge in the capacitor was zero.

Differentiating this expression with respect to time we obtain

$$R \frac{di(t)}{dt} + \frac{i(t)}{C} + L \frac{d i^2(t)}{dt^2} = 0 \tag{11.17}$$

which can be satisfied by a current

$$i(t) = A \exp(st) \tag{11.18}$$

where A is a complex constant whose value determines the current at the time $t = 0$ and

$$s = \frac{-R \pm \sqrt{R^2 - 4L/C}}{2L}. \tag{11.19}$$

If $R^2 \geqslant 4L/C$, s will be real and negative, and $i(t)$ will decay exponentially from its initial value $i(0)$. If $R^2 < 4L/C$, s will be complex. When this condition is satisfied, we can re-express s as

$$s = \alpha + j\omega \tag{11.20}$$

where

$$\alpha = \frac{-R}{2L} \qquad \omega = \frac{\sqrt{4L/C - R^2}}{2L}. \tag{11.21}$$

The current around the loop will then be

$$i(t) = A \exp(\alpha t) \exp(j\omega t). \tag{11.22}$$

This can be seen to indicate that the current variation with time takes the form of a sinusoidal oscillation at the frequency ω which decays exponentially (as α is negative) with a time constant $1/\alpha$.

Usually, we expect that R must be either positive or, perhaps, zero. However, if R were negative but still satisfies the conditions $R^2 < 2L/C$, then α would be positive and the oscillatory current would *increase* exponentially with increasing time. As a result, we can combine a negative-resistance device with a resonant circuit to produce an oscillator.

Clearly, the signal power produced by such an oscillator cannot appear from nowhere! The negative resistance must be supplied with *bias* power. From requiring that the oscillator must obey the law of energy conservation we may draw two general conclusions. Firstly, the mean output signal power must be less than or equal to the product of the applied bias voltage and current. Secondly, no device can have a negative resistance when unbiased, i.e. at 0 V or 0 A bias.

In some cases it is possible to employ an AC input to bias the device and it may then appear that the device has a negative resistance at a mean voltage and current of zero. In reality, however, the actual voltage and current in such a device is generally not zero. For most practical oscillators, the applied bias is a steady non-zero voltage and current.

As the device must exhibit a negative resistance at some appropriate non-zero applied voltage and current but must have a positive resistance at 0 V, it follows that real negative-resistance devices cannot obey Ohm's law. They hence fall into the category of *non-linear* devices. For a particular oscillator, the appropriate value of the negative resistance will be the *dynamic* resistance (the gradient $\delta V/\delta I$) at the chosen bias voltage or current.

As these devices have two terminals and display a non-Ohmic variation in current with voltage, they are frequently regarded as *diodes*. They are also sometimes called *active devices* to indicate their ability to transform power from a bias input into an output signal.

A considerable range of tuned circuits have been evolved for use in microwave and millimetre-wave oscillators. Figure 11.3 illustrates some typical arrangements for use with conventional rectangular waveguide. One of the most successful arrangements at millimetre-wave frequencies is the resonant cap circuit (figure 11.3(a)). In this system, an inductive post is placed across the waveguide and connects to the negative resistance device via a resonant *disc* or *cap*.

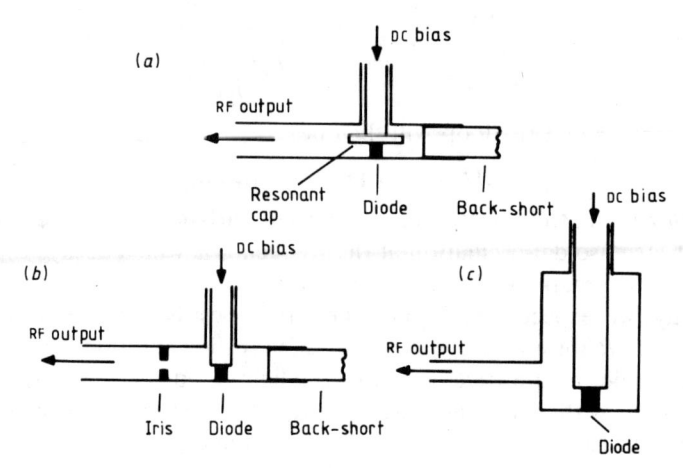

Figure 11.3 Typical waveguide arrangement for solid state millimetre-wave oscillators. (a) Resonant cap circuit. (b) Waveguide cavity. (c) Coaxial cavity.

Various types of active device may be used to make a millimetre-wave oscillator, each having its own particular characteristics. The choice of the optimum circuit arrangement will depend upon the properties of the device.

Although reliable measurements of the impedances of millimetre-wave oscillator diodes are scarce, theoretical considerations indicate that the magnitude of the negative resistance of most devices is of the order of an ohm. This is quite low compared with the characteristic impedances of most transmission lines which are typically around 100 Ω or more. The resonant circuit which defines the oscillation frequency must therefore also act as a *resonant transformer* to couple the generated power into the waveguide efficiently.

Figure 11.4(a) represents a simple tuned circuit and a back-short being used to couple a negative resistance $R = -g$ into a transmission line whose characteristic impedance is Z_0. At a signal frequency f the combination of the inductance L and the capacitance C with the negative resistance will produce an impedance $X(f)$ shunted across the transmission line, such that

$$X(f) = \frac{(X_L - g)X_C}{X_L - g + X_C} \qquad (11.23)$$

where

$$X_L = 2\pi f j L \qquad X_C = \frac{1}{2\pi f j C}. \qquad (11.24)$$

At most frequencies this impedance has both a real and an imaginary component; however, at the frequency f_0 which can be defined from the

Oscillators and Signal Sources

expression

$$(2\pi f_0)^2 \equiv \frac{L - Cg^2}{CL^2} \tag{11.25}$$

$X(f)$ is wholly real and we can rearrange the above to obtain

$$X(f_0) = -\frac{L}{C}\frac{1}{g} \tag{11.26}$$

i.e. at f_0, which is the *resonant frequency* of the circuit, the LCg combination appears as a negative resistance whose magnitude is proportional to $1/g$.

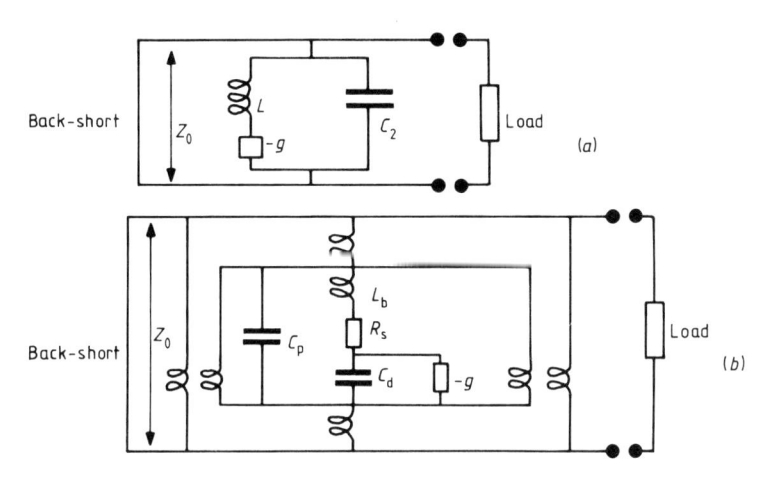

Figure 11.4 Electronic circuit analogues of a negative resistance device mounted in a resonant cavity. (a) Simple resonant circuit. (b) Realistic equivalent of 'cap' circuit.

The properties of a tuned circuit are often described in terms of a *quality factor Q* which can be defined by the expression

$$Q \equiv \frac{(2\pi f_0)L}{R}. \tag{11.27}$$

Combining this with equation (11.26) we can say that

$$X(f_0) = -g(Q^2 + 1). \tag{11.28}$$

For most cases of practical interest, $Q \gg 1$, and the above expression can be replaced by $X(f_0) \simeq -gQ^2$.

In low-frequency electronic circuits we can assemble a resonant circuit by using actual capacitances and inductances. At microwave and millimetre-wave frequencies, however, we must employ a resonator similar to the examples shown in figure 11.3. In such cases it is convenient to represent the properties of the circuit in terms of the quality factor Q rather than to attempt to define some appropriate values for L and C.

By reversing the arguments given above, we can see that, if the above circuit is connected across a matched transmission line of characteristic impedance Z_0, then the resistive load which terminates the guide is equivalent to a positive resistance

$$R \simeq \frac{Z_0}{Q^2} \qquad (11.29)$$

placed in series with the negative-resistance device. The combined effects of the device resistance $-g$ and this effective load resistance R upon the oscillator can therefore be represented by a total resistance $R - g$.

Using the arguments outlined earlier we can see that oscillation at the signal frequency f_0 can only take place provided that

$$0 \geqslant R - g \geqslant -\sqrt{\frac{2L}{C}}. \qquad (11.30)$$

Now, if $R - g$ is negative, we may expect the magnitude of the oscillation to grow exponentially. Obviously in a real oscillator the oscillatory current cannot continue to grow indefinitely, and so we can expect that some process will eventually act to limit any further increase in the magnitude of the oscillation.

It has already been pointed out that a real negative-resistance device cannot have a negative resistance at 0 V. In order to function as an oscillator, some bias must be applied in order to provide a mean device voltage and current at which the dynamic resistance $\delta V/\delta I$ is negative. Once oscillation has begun, however, the device voltage and current will vary periodically about this mean value. As the magnitude of oscillation increases, the device voltage and current will sweep back and forth over an increasing range until, eventually, it will enter regions where $R - \delta V/\delta I$ is no longer negative.

Whilst $R - \delta V/\delta I$ is less than zero, any oscillatory energy in the circuit will tend to increase as time passes. However, whilst $R - \delta V/\delta I$ is greater than zero, any oscillatory energy will tend to be dissipated by the positive resistance. If we define g to be the value of $\delta V/\delta I$ averaged over each cycle of oscillation, we may see that the magnitude of the oscillatory signal will rise to a stable equilibrium at which $R - g = 0$.

Any further increase in the magnitude of oscillation will tend to increase the amount of time that the device spends outside the *negative-resistance region*—the range of voltages where $R - \delta V/\delta I$ is negative—and cause more power to be dissipated than is being generated. This process hence limits the magnitude of the oscillation. Once a stable oscillation has been reached, the total resistance around the circuit (including both the device and the effective load resistance) will be zero. The oscillation frequency f_0 will then simply be the resonant frequency

$$f_0 = \frac{1}{2\pi\sqrt{LC}}. \quad (11.31)$$

In practice the properties of most millimetre-wave oscillator systems are more complicated than the simple form of tuned circuit shown in figure 11.4(a). In order to take this into account, we may devise a circuit which models more reliably the detailed performance of a particular system.

Figure 11.4(b) illustrates a typical circuit model of an oscillator which uses a resonant cap and inductive post to couple a negative-resistance device into a conventional rectangular metallic waveguide. In addition to the device negative resistance $-g$, the model also takes into account a number of other factors including the device capacitance C_d, the inductance L_b and resistance R of the wires connecting the device to the top of its package and the package capacitance C_p.

We may regard this sytem as being composed of two parts: a device of impedance $Z_d = -g(f, A) + jX_d(f, A)$ and a circuit which presents an impedance $Z_c = R(f) + jX_c(f)$ to the device terminals, where (f) indicates dependence upon signal frequency and (A) dependence upon the signal magnitude. This 'total' circuit impedance Z_c takes everything external to the device into account, including the load and back-short impedances which terminate each end of the transmission line.

Although the detailed behaviour of such a system is more complicated to evaluate in detail, its basic properties are much the same as the simple example considered earlier. These may be summarized by saying that we expect the magnitude of the oscillation to be such that

$$R(f_0) - g(f_0, A) = 0 \quad (11.32)$$

and the oscillation frequency f_0 to be such that

$$X_d(f_0, A) + X_c(f_0) = 0. \quad (11.33)$$

Both the real and the imaginary parts of Z_d generally depend upon the bias applied to the device; hence we can expect the magnitude and frequency of the oscillation to depend upon the bias. Any change in the back-short position and the load impedance will also affect the oscillator output. The effect of a mismatched load is generally referred to as *load pulling* as it reflects power back onto the negative resistance device and 'pulls' the oscillation frequency away from the value that it would have if the load matched the transmission line impedance.

Although at first examination the combination of negative resistance and a resonant circuit appears very different from that of an amplifier and a feedback network, the two arrangements have an underlying similarity. In order to make this clearer, we can consider a simple arrangement of negative resistance $-g$ in series with a positive resistance R.

If we apply an input voltage V_i across the pair of resistors, the resulting current would be

$$I = \frac{V_i}{R - g} \tag{11.34}$$

and the input power level P_i would be

$$P_i = I^2(R - g). \tag{11.35}$$

However, the voltage V across the positive resistance will be

$$V = IR \tag{11.36}$$

and the power P dissipated by the resistance R will be

$$P = I^2 R. \tag{11.37}$$

Combining these expressions we find that

$$P = P_i + gI^2 \tag{11.38}$$

i.e. the power dissipated in the resistance R is *greater* than that apparently input into the circuit by an amount which depends upon the magnitude of the negative resistance $-g$. Hence we can regard the negative resistance as producing a *power gain*

$$G = \frac{R}{R - g}. \tag{11.39}$$

The power dissipated in the load R can then be written as

$$P = GP_i. \tag{11.40}$$

The negative-resistance device hence acts as a signal amplifier whose power gain is G.

A resonant circuit stores energy at the frequency of oscillation. This energy can be regarded as being transferred back and forth between the electric and magnetic fields within the cavity (or within the capacitance and around the inductance). Any change in the input signal voltage will produce a similar change after a time delay corresponding to one cycle at the resonant frequency. Hence the resonator can be seen as a form of feedback circuit which returns output signals after a suitable delay.

From this viewpoint we can see that there is a fundamental similarity between the amplifier–feedback and negative-resistance–resonator forms of oscillator.

In general it is useful to employ an oscillator circuit which has a high value of the quality factor Q. This will help to minimize the frequency shifts induced by a given level of residual signal reflection. It is also valuable in ensuring that the oscillator output is as free as possible of sideband noise.

Random processes such as thermal noise, shot noise, etc, within the negative-resistance device will cause fluctuations in the oscillatory current or voltage. It should also be noted that, although on average $R - g = 0$ for a stable oscillation, there will be instants during each cycle when $R - g$ is negative—even for signal frequencies which differ slightly from f_0. Hence small levels of noise at these frequencies may be amplified by the negative resistance. As a result, the output signal will not simply consist of a pure sinusoid at the resonant frequency f_0.

At frequencies very near f_0 the resonant circuit will continue to behave as an effective transformer and any noise generated can be coupled to the output load with a reasonable degree of efficiency. For frequencies well outside the range from $f_0 + \delta f$ to $f_0 - \delta f$, where $\delta f = f_0/Q$, the resonant circuit will not efficiently match any noise power into the output load.

The real part $\mathrm{Re}(f)$ of the circuit impedance presented to the negative-resistance device will only be comparable in magnitude with $-g$ over the range $f_0 + \delta f > f > f_0 - \delta f$. As a result, any tendency for noise to be amplified will be restricted to a range of signal frequencies which will depend upon Q. The value of the factor Q is therefore important in determining both the amount of unwanted noise which the oscillator may produce and the efficiency with which noise is communicated to the load.

11.4 Solid State Negative-resistance Devices

Here we shall outline the basic properties of three types of negative-resistance device which can be used in millimetre-wave oscillators: the *impatt*, *Gunn* and *double-barrier diodes*.

11.4.1 The impatt diode

The word *impatt* is formed as an abbreviation of *imp*act *a*valanche *t*ransit *t*ime and serves as a mnemonic for the method of operation of the device. Various types of semiconductor structures have been devised which operate as impatt devices. Although the complexity of these structures and their detailed performance vary, each employs the same basic method of operation. For example we can consider one of the simplest, and earliest, types of impatt structure, illustrated in figure 11.5(*a*). This structure is sometimes referred to as a *Read diode*.

The device essentially consists of a p–n junction connected to a region of intrinsic semiconductor. End contacts can then be used to apply a bias voltage to the device. In normal operation the applied voltage produces a *reverse bias* across the p–n junction, i.e. the voltage widens the carrier-depleted region near the junction and virtually no current can flow through the device.

Although the p–n junction is reverse biased, the current flowing through the device is not zero. Thermal effects will generate a small number of carrier electron–hole pairs in the depletion region which are swept apart by the high electric field. This current will depend upon the applied voltage and the device temperature.

For moderate applied voltages the electric field distribution along the device is of the general form shown in figure 11.5(b). Throughout most of the device some carriers are available and the conductivity is much higher than in the depletion region. Hence nearly all the applied potential difference appears as a large electric field near the p–n junction.

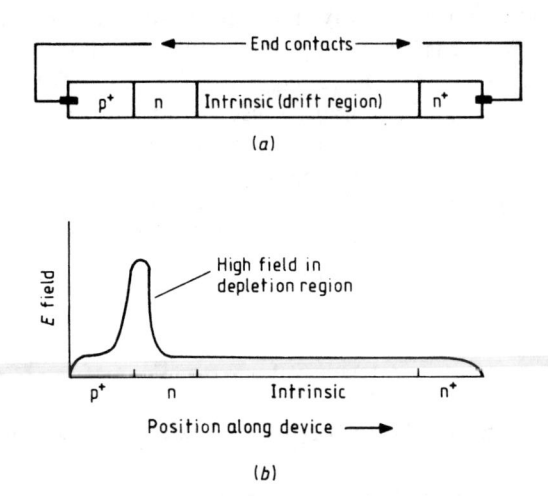

Figure 11.5 (a) Physical model of a 'Read diode' impatt device. (b) Static electric field along an impatt, biased below threshold.

Carriers tend to be accelerated by the applied field until they are slowed down by a collision. This process then repeats itself and, as a result, the carriers tend to move along the device with a mean velocity determined by the electric field and the average distance between collisions. If, however, the applied electric field is very high, then the carrier may gain sufficient energy to ionize an atom with which it collides. This produces a new carrier pair which becomes part of the current flowing through the device. The new carriers produced by *impact ionization* can also be accelerated and produce further carriers.

As a result, when the applied field exceeds a particular level, the current will begin to rise rapidly. The rate of current increase will then depend both upon the voltage *and* upon the existing current. This process is an *avalanche* of charge carriers and causes the current to rise rapidly even if the applied electric field is kept constant. The voltage at which a current avalanche may begin is called the *avalanche threshold voltage*.

To see how an impatt device acts as a negative resistance, we can assume that the total field across the device at any moment is proportional to a

Oscillators and Signal Sources

combination of two applied potentials: a steady bias V_T set at the avalanche threshold level, and an alternating signal V_0 which varies the total above and below this threshold.

Figure 11.6 illustrates the behaviour of an impatt device under these conditions. Since the rate of carrier generation depends upon both the applied field and the number of carriers already available, the current does not initially rise very quickly when $V_0 + V_T$ moves above the threshold value. (It also takes a short, but finite, time to accelerate 'new' carriers and to give them enough energy to create further carriers.) After a short delay, however, the rate of carrier production begins to rise with increasing rapidity.

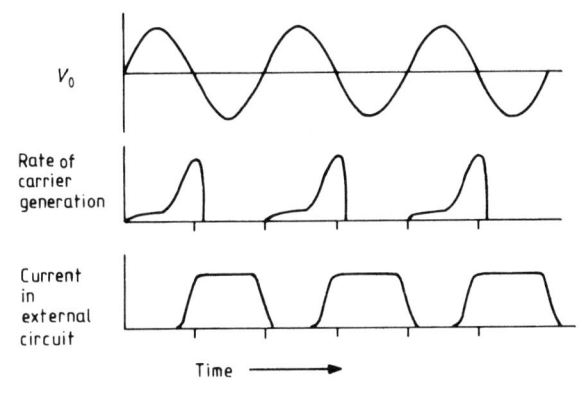

Figure 11.6 Dynamic behaviour of an impatt.

This process of carrier production will continue until the total field falls below threshold or until no more potential carriers are available. For the case shown in figure 11.6 we assume that the applied field falls below threshold after a given time and carrier production abruptly comes to an almost total halt.

The electric field within the device is largely concentrated near the p–n junction. Hence the brief period of impact ionization has effectively produced a bunch of electrons near the junction. This bunched charge now moves along the device with a mean velocity determined by the properties of the intrinsic semiconductor and the level of the applied field. The length within the device which the charge has to traverse is called the *drift distance*.

As the bunched charge moves away from one end contact and towards the other, its surrounding field causes the potentials at the contacts to change. The contact that it is approaching becomes more negative and the one that it moves away from becomes more positive. Hence, whilst the charge bunch is drifting through the device, a current will tend to flow in any external circuit which links the end contacts.

The time delays associated with carrier generation and passage through the drift region mean that the current in the external circuit will flow for a particular time after the brief increase in applied voltage which set the process in train. An applied periodic voltage variation of the correct frequency will therefore, as shown in figure 11.6, produce current bursts which are essentially 180° out of phase with the voltage variation. Hence, at that frequency, the device behaves as if it has a negative resistance.

The levels of electric field required to initiate the impact ionization avalanche process are quite high. A typical silicon impatt device may have a depletion layer width of the order of a micron and an avalanche threshold voltage of around 10–20 V. The mean current density in a working device will generally be around 10 000 A cm^2. As a result, the operating temperatures of impatt devices also tend to be high.

Silicon impatt devices have proved able to provide relatively high continuous power levels (compared with Gunn devices) at frequencies up to around 200 GHz where between 50 and 100 mW can be obtained. Impatt devices do, however, suffer from some disadvantages which often outweigh the high powers that they offer.

Impatt devices rely upon an avalanche effect which is inherently random (noise like) in nature. This means that impatt devices tend to produce more sideband noise than Gunn devices do. The avalanche mechanism also means that, unless prevented, the current increase which starts when threshold is exceeded may continue until the high current and power dissipation destroy the device. For the explanation given above, it is assumed that the field across the device is already alternating about the threshold value at an appropriate frequency. If, however, the field remains above threshold for too long, then the device may self-destruct.

In order to help to prevent the device from destroying itself, impatts are generally provided with bias power from a constant-current supply. Any tendency for the device to draw too much current then causes the applied voltage to fall, limiting the current to a safe value. The use of this form of bias supply also helps to take into the account the tendency of the avalanche threshold voltage to alter as the device temperature changes.

At a junction temperature of 250 °C an impatt device can be expected to operate continuously for over 100 000 h. Used with care, therefore, millimetre-wave impatts offer a longer service life than do vacuum state oscillators. However, impatts are prone to failure when being switched on or off.

This is partly due to the possibility that the initial current in the device may be too high, causing the device to fail before the desired oscillation can establish itself. For the same reason, impatt devices may fail if connected to an unmatched load or if the cavity is adjusted without care. Failure may also be produced by the abrupt change in temperature which takes place when the bias is switched on. Any interference 'spikes' which appear on the bias supply to the device may also lead to abrupt failure.

In the type of impatt device described above, it is assumed that only one type of carrier—electrons—drifts through the device and hence it is referred to as a *single-drift* device. An alternative form of impatt places the p–n junction between a pair of drift regions. Impact ionization creates both electrons and holes (n- and p-type charge carriers). The two types of carrier then move apart through the opposing drift regions. Impatt devices of this type, referred to as *double-drift* devices, tend to produce more power than single-drift devices do.

11.4.2. The Gunn diode

The *Gunn diode* (or *transferred-electron device*) offers a much higher level of device reliability, and a generally lower level of sideband noise. The only significant disadvantage of the Gunn device compared with the impatt is a lower continuous power level. The device operates by exploiting an effect observed by J B Gunn.

Most devices referred to as diodes make use of effects which arise when two materials with differing conduction properties are placed together forming a junction. The Gunn device is based upon a bulk property of a suitable semiconductor. In principle, therefore, it may simply consist of a uniform piece of material with a pair of Ohmic end contacts. A device of this type shows a variation in current with applied voltage which does not change its form if the polarity of the applied voltage is reversed.

In practice the performance of Gunn devices can be significantly enhanced by producing inhomogeneous devices and hence real Gunn diodes do not usually display symmetric electrical properties.

We can relate the conductivity σ of a material to the density n_c of available carriers and their mobility μ_c. An applied field E will produce a current density J by

$$J = \sigma E = n_c q \mu_c E \tag{11.41}$$

where $q = 1.6 \times 10^{-19}$ C, the charge on a single carrier.

At low levels of the applied electric field, essentially all the charge carriers will be in the lowest available conduction band. In GaAs, electrons in this band have a low-field mobility μ_c of about 8000 $cm^2 V^{-1} s^{-1}$, i.e. for an applied field of 1 $V cm^{-1}$ in GaAs material where $n_c \simeq 10^{16}$ cm^{-3}, the current density J would be around 12.8 $A cm^{-2}$.

In GaAs, another band exists which is around 0.35 eV above the lowest-energy conduction band. This band is usually almost empty but, if the applied field is increased sufficiently, carriers can gain enough energy to make it possible to transfer to this higher band. The field required to achieve this effect is called the *threshold* field. For fields up to about 3.5 $kV cm^{-1}$, most carriers remain in the lower band and the current density rises essentially in proportion to the applied field. Above these levels, however, a significant fraction of the carriers are able to transfer to the higher conduction band.

It is conventional to refer to the lower conduction band as the *central* Γ *valley* and the upper band as the *satellite* L *valley*. When plotted on a graph of carrier energy against momentum, these two bands appear as roughly parabolic curves. The lower band is symmetric about zero momentum (hence central), and the upper is displaced to one side (hence satellite).

Carriers in the higher band have a higher effective mass and a significantly lower mobility than in the lowest-energy conduction band. As a result, there is a tendency for the current density to *fall* as the increasing applied field transfers more and more carriers to the higher band. As a consequence the material exhibits a negative resistance for applied fields in excess of $3.5\,\text{kV}\,\text{cm}^{-1}$.

Although most Gunn devices manufactured to date have been made using GaAs, other semiconductor materials show the same effect. InP, for example, has a satellite band with a lower mobility 0.53 eV above the main conduction band. Significant numbers of carriers can be transferred to this band if the applied field is around $10\,\text{kV}\,\text{cm}^{-1}$ or more.

Gunn devices operate at much lower levels of applied field than impatts do and do not employ an avalanche process which may run out of control if the applied field stays above threshold for too long. If the field applied to a Gunn device is too high, then the device tends to settle into its 'low-mobility' state at an appropriate steady current. The device can usually be engineered to endure this state for a reasonably long period. Hence Gunn devices are more reliable than impatts and are robust against bias or load problems. The absence of the avalanche mechanism also means that well designed Gunn oscillators can offer low levels of sideband noise.

At high frequencies, the ability of Gunn devices to show a negative resistance tends to be limited by the time taken for carriers to change from one band to another. Carriers in the lower band are accelerated by the applied field and will take a finite time to reach the required energy. Each carrier must then move through the material until it interacts (i.e. collides with something), which allows the momentum change required for transfer to the satellite band. The time taken for carrier transfer hence depends upon the applied field and the properties of the chosen material.

Having obtained a significant carrier population in the upper band we find that some time will elapse following a reduction in the applied field before many carriers return to the lower band. This is because of the finite time that it will take, on average, until the carriers become involved in an interaction which causes their energy and momentum to change, transferring them back to the lower band.

The typical time t taken to accelerate a carrier up to the required transfer energy $\mathscr{E}$ can be obtained from

Oscillators and Signal Sources

$$t^2 = \frac{2\mathcal{E}m}{E^2 e} \tag{11.42}$$

where m is the mass of the carrier, e is its charge and E is the applied field.

The threshold field in InP is around three times that in GaAs; hence, even though the carriers must attain more extra energy, they do so relatively swiftly. An applied field of 3.5 kV cm^{-1} will take around 5.7 ps to accelerate the carriers in GaAs to the required transfer energy. In InP the higher applied field of 10 kV cm^{-1} will accelerate the carriers up to the required energy in about 2.4 ps. The inertial energy relaxation time—a measure of how long a carrier must wait before becoming involved in an interaction—is around 0.75 ps in InP compared with 1.5 ps in GaAs.

As a general rule of thumb we can estimate the highest possible oscillation frequency by adding together the acceleration and interaction delays and taking the result as being the time required for one half-cycle of the oscillation. This implies that GaAs Gunn devices would oscillate at frequencies up to about 70 GHz, and InP up to around 160 GHz. In practice, the high-frequency behaviour of devices is more complex than this assumes, but a more careful analysis yields upper frequency limits reasonably close to these values.

The high-frequency output from Gunn devices can sometimes be extended by employing modified devices or modes of operation. For example, it is possible to form cathode contacts which launch 'hot' carriers into the device, i.e. carriers which enter already having a large part of the energy required for transfer to the upper band. This can produce a marked improvement in the high-frequency performance by getting rid of the 'dead region' near the cathode which carriers must otherwise cross whilst gaining the required energy.

Many millimetre-wave Gunn oscillators also employ a *harmonic* method of operation. The device is constructed so as to work efficiently at half (or sometimes a third) of the required frequency. The device then acts as its own non-linear frequency multiplier, converting some of the lower-frequency power into harmonic signals which can be extracted. Most GaAs Gunn oscillators used at frequencies above around 80 GHz operate in this way.

The maximum power output from a Gunn device is limited by its physical size, the input bias power level and the efficiency with which input bias power is converted to oscillatory output. The size of a device can be characterized in terms of its *active* length (the length of material over which the Gunn effect is used to produce a negative resistance) and its cross sectional area (which relates the signal current to a particular current density.

As the required signal frequency increases, it is generally advisable to reduce the active length. If this is not done, problems may arise owing to a non-uniform field and current distribution within the device. When the *transit time* (the time taken for signal current carriers to pass through the active region) becomes a large fraction of one cycle period, *domains* may begin to form. Here the field within the device develops bunches of carriers which are in the low-mobility band and move relatively slowly through the device. The faster carriers in front of such a region will move away, and those behind will be accelerated into the low-mobility bunch. This tends to cause the applied field to peak also at the bunched 'domain', further increasing the tendency of the low-mobility carriers to form a bunch.

Eventually, the field at the domain may rise to the point where the device material is damaged. At lower fields the domains may form erratically from one cycle to another, producing noise and instability problems. The formation of domains can, if controlled with care, produce efficient oscillators at low signal frequencies. At high frequencies, however, this method of operation is usually avoided by keeping the device short. This simplifies operation and provides a quiet reliable source.

Any input bias power which does not appear as output oscillatory signal is wasted as resistive losses within the device. Hence there is a tendency for the device to become hot in use. A short active length eases the problem of heat sinking the device. In practice, a device which is too long will overheat because the end farthest from the heat sink must be cooled by thermal conduction through the length of the device.

The main disadvantage of a short active length is that it will tend to increase the shunt capacitance between the ends of the device. This is a particular problem at high frequencies as it may produce a device which is virtually shorted out by its own capacitance. To prevent this, the device cross sectional area must be reduced. As a result both the active length and the cross sectional area of millimetre-wave Gunn devices tend to be reduced as the required oscillation frequency rises. For given values of threshold field and current density this means that the applied bias voltage, current and power must fall as the required frequency rises.

The efficiency with which input bias power can be converted into output oscillations is limited by the change in conductivity which takes place when carriers are transferred from one band to another and the change in applied field required for most carriers to change band. In practice, this is generally represented in terms of the *peak-to-valley ratio k*. The mean carrier velocity can be measured (or predicted) as a function of the applied field. The peak-to-valley ratio can then be defined as the maximum value of the mean velocity divided by the minimum value obtained at a higher applied field. For GaAs, $k \simeq 2.2$ and, for InP, $k \simeq 3.5$. Hence we can expect InP to provide more efficient Gunn oscillators than GaAs does.

GaAs devices have produced continuous power levels of around 50 mW at 80 GHz and a few milliwatts at frequencies above 120 GHz. It is not always clear from reported measurements whether a particular device is

Oscillators and Signal Sources

operating as a fundamental or a harmonic oscillator. Over the 80–200 GHz region it can be expected that InP devices will provide more power than GaAs devices. Results obtained to date have shown output power levels in excess of 100 mW at 80–90 GHz. Theoretical estimates based upon the properties of InP indicate that around 80 mW should be possible from an InP device at 140 GHz. In excess of 10 mW should be possible from a harmonic InP Gunn oscillator at 200 GHz.

11.4.3 Multiple-barrier or quantum well devices

During the last decade or so, semiconductor fabrication techniques have advanced to the point where it has become possible to manufacture devices whose properties depend upon complex structures whose dimensions are comparable with—or smaller than—the typical mean free path of carriers moving within the device. For this reason, structures of these sorts are sometimes referred to as *low-dimensional structures* and the devices as *low-dimensional devices*. Various types of device have been considered, but here we shall only examine one simple type—the *double-barrier diode*—and its use in millimetre-wave oscillators.

The detailed behaviour of a double-barrier diode is quite involved and is currently the subject of much debate. As a result it is not yet clear just what the ultimate performance limitations of this type of device will be. At the time of writing, only a few working oscillators have been made and tested. It does, however, seem that low-dimensional devices of this—or a related type—will prove useful in making solid state oscillators at frequencies significantly higher than those available from GaAs or InP Gunn oscillators.

For simplicity it is convenient here to make use of an analogy to the Fabry–Perot resonator in order to explain the main properties of a double-barrier diode. This enables us to establish how the device may exhibit a negative resistance.

Figure 11.7 illustrates a practical structure taken from the current literature which consists of two thin layers of undoped AlAs which are separated by a layer of undoped GaAs. At either side of the structure the material is GaAs, n-doped to a concentration of around 2×10^{17} cm^{-3}. The AlAs layers are 1.7 nm thick and the undoped GaAs layer 4.5 nm thick.

The electron conduction band in the AlAs is at a considerably higher energy than in the GaAs; hence the AlAs layers are depleted of electrons and act as barriers through which carriers can only normally pass by tunnelling. The inner undoped GaAs layer can be regarded as a 'leaky' potential well, i.e. any carriers placed in this layer will tend to be confined by the AlAs walls but may eventually escape by tunnelling.

The properties of an electron moving through the device can be described in terms of a wave whose frequency and wavelength are related to its energy and momentum. At any specific moment, this wave will be coherent over a distance called the *coherence length* related to the mean free path length of an electron. An electron arriving at a barrier will be either reflected or transmitted. When considered in terms of its wavefunction, the probability of reflection or transmission can be obtained from the relative magnitudes of the reflected and transmitted components of an incident wave.

If the barriers are so far apart that they are separated by more than the coherence length of the electron wave, then the behaviour of an electron arriving at one barrier will not be affected by the existence of the other. Under these circumstances, the propagation of the electron wave can be treated by considering the effect of each barrier it encounters in turn.

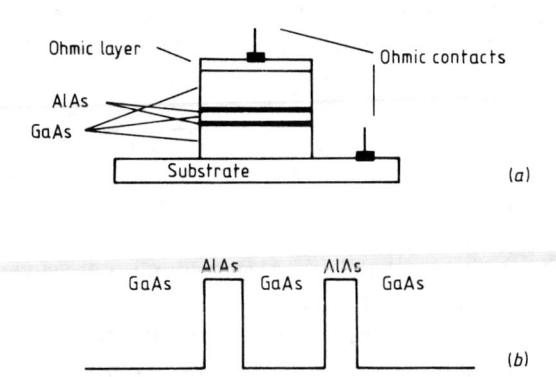

Figure 11.7 Double-barrier diode structure. (a) Schematic diagram and (b) structure of a double-barrier diode.

When the distance between the barriers is small compared with the coherence length, an incident electron will produce waves at the two barriers which are coherent with one another. As a result, the wave components reflected by the barriers will have a specific phase relationship. The total reflectivity of the pair of barriers now depends upon the distance between the barriers compared with the wavelength of the electron wave.

When the distance between the barriers is such that the reflected components share the same phase, then the magnitude of the total reflected wave is greater than that produced by a single barrier. When the distance is such that the reflected components are 180° out of phase, then the magnitude of the total reflected wave is reduced. For two identical thin barriers we shall therefore observe a maximum reflectivity (and hence minimum transmissivity when the distance between the barriers is $\frac{1}{2}n\lambda$, where n is an integer and λ the electron wavelength. Similarly, the reflectivity will be at a minimum when the distance between barriers is $\lambda(\frac{1}{2}n + \frac{1}{4})$.

The wavelength of the incident electrons will depend upon their energy which, in turn, will be determined to some extent by the applied field. The conductivity of the device will depend upon the ease with which carriers can pass the barriers. As the potential difference applied between the ends of the device is increased, the energy of arriving carriers rises and their wavelength falls. Hence we find that the device conductivity tends to rise and fall alternately with increasing applied voltage as the electron wavelength changes.

The behaviour of the double-barrier diode is analogous to a Fabry–Perot resonator illuminated with photons of a given wavelength. As a result the tendency of the device conductivity to peak at specific levels of applied field is referred to as *resonant tunnelling*. At an applied field level just above that which produces a resonant peak the conductivity will tend to fall with increasing applied field, i.e. the device exhibits a negative resistance when biased to such a field.

In order to establish the highest possible frequency of oscillation, we can make use of a general quantum mechanical argument and link the response time to an energy uncertainty. If we calculate or measure how the device conductivity (or device transmissivity) varies with the carrier energy, we can establish that the resonant transmission peak will have a 'width' $\Delta \mathscr{E}$, i.e. the conductivity will be high over an energy range $\Delta \mathscr{E}$ centred upon the peak value. This energy can then be regarded as an uncertainty in the energy required for tunnelling to occur and implies a response time

$$\tau = \frac{h}{2\pi \Delta \mathscr{E}}. \qquad (11.43)$$

The highest possible oscillation frequency can then be estimated to be around $1/2\tau$.

The above method for estimating the device response time is appealing in the context of an analogy to the Fabry–Perot resonator as we find that the time taken, at resonance, for the reflected or transmitted light level to alter in response to a change in illumination is related to the width Δf, of the resonant peak, in a similar way. Despite this convincing parallel, however, it is important to remember that an analogy is *not* an identity, i.e. the apparent similarity may prove misleading. The behaviour of charged carriers in double-barrier devices differs in detail from that of photons in a Fabry–Perot resonator.

An analysis of the type of device described above leads to a value of τ which implies a maximum possible frequency of perhaps 600–800 GHz. If the AlAs barriers were to be replaced by $Ga_{1-x}Al_xAs$, then it becomes theoretically possible to obtain values of τ which imply oscillation frequencies around 2000 GHz. Hence it appears that, by an appropriate choice of structure dimensions, materials, etc, solid state oscillators might be operated at very high millimetre-wave frequencies.

Unfortunately, this simple approach fails to take some problems into account. For example, the response of an actual device may be limited by the transit time required for carriers to travel the length of the device. Also, we may find that the shunt capacitance of the device prevents operation at very high frequencies. The maximum practical oscillation frequencies of these double-barrier devices will as a consequence most probably be in the 200–400 GHz region.

In order to obtain high operating frequencies a structure must be chosen which maximizes $\Delta\mathscr{E}$. This tends to lower the peak-to-valley ratio, restricting the oscillator efficiency. Furthermore, the device must be small in order to minimize transit and capacitance problems. As a result the power output from simple double-barrier devices is liable to be small at high frequencies.

It may well prove that alternative types of device will yield better power output or higher practical oscillation frequencies. One possible candidate for an improved structure has been suggested where a double-barrier arrangement is coupled to a drift region. The resulting device can be analysed in a manner similar to an impatt device. However, the avalanche process which injects carriers into the drift region of an impatt is replaced with the negative resistance of the double barrier. This means that the device should not share the noise and reliability problems of a conventional impatt device.

Chapter 12

Frequency Control Loops and Diplexers

Figure 12.1 illustrates the arrangement of a typical millimetre-wave heterodyne receiver system. In this chapter we shall examine the two main parts of such a system which have not already been considered in other chapters: the control loop which determines the local oscillator (LO) frequency, and the diplexer which simultaneously couples LO and signal power into the mixer.

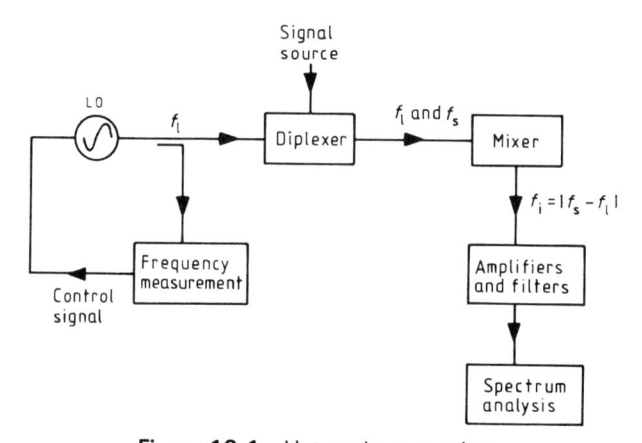

Figure 12.1 Heterodyne receiver.

12.1 Local Oscillator Control

In most practical systems we require an LO whose output frequency is well known. Any uncertainty in the LO frequency will permit errors to arise when we attempt to calculate the signal frequency from measurements of the output intermediate frequency. Control of the LO frequency is particularly useful for dealing with problems such as thermal drift or sideband noise which will arise with any real oscillator.

In order to be able to control the LO we must be able to measure its oscillation frequency and generate a *control signal* which can be fed back to the oscillator so as to correct its output in the desired way. A variety of detailed arrangements have been devised to carry out this task. Many of these can be seen to be similar to the arrangement shown in figure 12.2 which represents a system which employs a *harmonic mixer*.

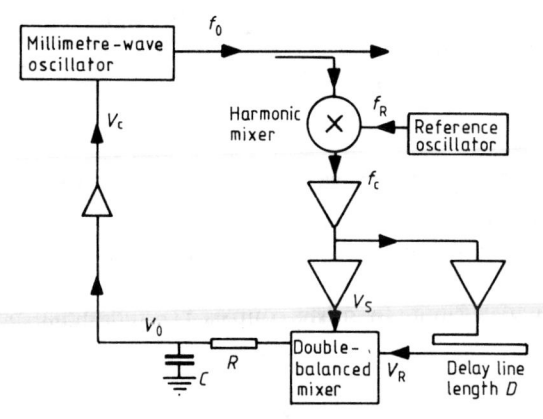

Figure 12.2 Frequency control loop for defining the LO frequency supplied to the mixer in a heterodyne system.

A fraction of the LO output power is injected into a harmonic mixer along with the output from a *reference oscillator*. In a typical system, the reference oscillator frequency f_R will be chosen to fall in the 1–10 GHz range. The reference oscillator input will tend to induce oscillatory currents in the mixer at its harmonic frequencies nf_R. When some of the LO power, at a frequency f_0, is also coupled into the harmonic mixer, it becomes possible to obtain an output from the harmonic mixer at one (or more) of the frequencies $f_c = |f_0 - nf_R|$. Provided that f_R and the harmonic number n are known, we can now use f_c in place of the higher frequency f_0 as any change in f_0 will also alter f_c by the same amount.

This method of using a harmonic mixer is convenient in that f_c can be arranged to be much lower than f_0. In many practical systems it will be arranged that the system provides a value of f_c in the region from 100 MHz to 1 or 2 GHz where it is generally possible to obtain electronic components of excellent performance. The main advantage of this technique can, however, be seen by considering a simple example.

Consider a system where the desired LO frequency f_0 is 100 GHz using a reference oscillator frequency f_R of 5.005 GHz. We can expect that the

Frequency Control Loops and Diplexers

harmonic mixer will provide some output at the frequency $f_c = 100$ MHz as a result of mixing f_0 with nf_R where $n = 20$. If the actual LO frequency is in error by 1 MHz, then the fractional change in f_0 will be 1 part in 100 000. Such a small fractional change would prove difficult to measure directly by, say, the use of an optical resonator. However, this error in f_0 will also cause f_c to change by 1 MHz, i.e. the fractional change in f_c would be 1 part in 100, a change which would be relatively easy to measure.

By employing the harmonic mixer as a frequency converter we can essentially magnify small fractional changes in f_0 into quite large fractional changes in f_c. This makes it possible to detect changes in f_0 of 1 part in a million or less without difficulty. Hence we can hope to sense and correct very small errors in f_0.

Having obtained the output f_c, we must now measure this frequency to obtain a control signal which can be used to correct f_0 if it departs from the desired value. In the arrangement shown in figure 12.2 the frequency f_c is measured using a *double-balanced mixer* (DBM) in conjunction with a *delay line*.

Figure 12.3 illustrates the action of a DBM. The circuit consists of a set of four diodes used as connections between two transformers. Each of the transformers has a centre tap on one of its windings. Any *reference signal* voltage V_R applied between the centre taps will induce equal and opposite current in each half-winding and hence will not produce any output via either transformer. Provided, however, that $|V_R|$ is high enough, the reference signal will cause one pair of diodes to conduct (i.e. have a low dynamic resistance) and the others to become reverse biased (i.e. have a high dynamic resistance).

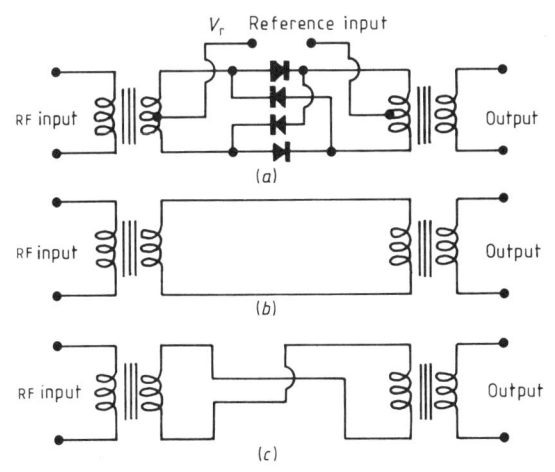

Figure 12.3 Double-balanced mixer: (a) circuit; (b) equivalent circuit when $V_r > 0$; (c) equivalent circuit when $V_r < 0$.

The arrangement will therefore behave as if the input and output transformers were connected as illustrated in figure 12.3(*b*) or 12.3(*c*) depending upon the sign of V_R. Hence we can arrange that, for example, when $V_R > 0$, the signal input is transmitted unaffected to the output. When $V_R < 0$, the connections will be reversed and the input signal will be *inverted* on its way to the output. The packaging, documentation, etc, for most DBMs will refer to the signal and reference inputs as the RF and LO ports and the output as the IF *port*. This usage has been avoided here to avoid confusion.

The DBM acts in much the same way as a *phase-sensitive detector*. If we apply a reference signal

$$V_R = A \cos(\omega t) \tag{12.1}$$

and an input

$$V_I = B \cos(\omega t + \phi) \tag{12.2}$$

then the *mean* output voltage V_0 averaged over a time much greater than $1/\omega$ will be

$$V_0 = B \cos \phi. \tag{12.3}$$

Hence, by employing a suitable output filter, we can obtain a voltage V_0 which depends only upon the magnitude B of the input and its phase ϕ compared with the reference signal.

In the system shown in figure 12.2, the reference signal V_R is identical with the input signal V_s except for having passed through a delay D; hence we can say that

$$V_s = B \cos(2\pi f_c t) \tag{12.4}$$

$$V_R = A \cos\left(2\pi f_c t + \frac{2\pi f_c D}{c}\right) \tag{12.5}$$

where c is the velocity of the signal along the delay line. From these expressions it follows that

$$V_0 = B \cos\left(\frac{2\pi f_c D}{c}\right) \tag{12.6}$$

i.e. the output voltage varies sinusoidally with f_c.

If the system is set up such that the desired LO frequency f_0 produces from the harmonic mixer an output frequency

$$f_c = \left(\frac{c(n + \tfrac{1}{2})}{2D}\right) \tag{12.7}$$

then the reference and signal inputs to the DBM will be 90° out of phase and V_0 will be zero when f_0 has the required value. If, however, the oscillator

Frequency Control Loops and Diplexers

frequency shifts to $f_0 + \Delta f$, then the output voltage will become

$$V_0 = B \sin\left(\frac{2\pi \Delta f D}{c}\right) \qquad (12.8)$$

which, provided that $\Delta f \ll c/4D$, is approximately equal to

$$V_0 = \frac{B 2\pi \Delta f D}{c}. \qquad (12.9)$$

The combination of the DBM and the delay line acts as a *frequency discriminator* to provide an output signal whose magnitude and sign is a measure of the frequency change Δf. V_0 can therefore be amplified to provide a control voltage V_c which can be used to alter the frequency generated by the millimetre-wave oscillator back towards the required value.

In practice, a frequency control loop of this type cannot completely correct for any frequency errors although it may significantly reduce their magnitude. Consider a simple example of an oscillator whose output frequency is both temperature and bias voltage dependent. The oscillator output frequency f' can be defined from

$$f' = f_0 + \frac{\delta f}{\delta T}(T' - T) + \frac{\delta f}{\delta V}(V' - V) \qquad (12.10)$$

where T and V are the nominal operating temperature and bias voltage, and T' and V' are the actual temperature and voltage. A change ΔT in the oscillator temperature will—if the bias voltage is fixed—change the output frequency by an amount

$$\Delta f(T) = \frac{\delta f}{\delta T} \Delta T. \qquad (12.11)$$

For a given amount of LO power injected into the harmonic mixer we shall obtain an output

$$V = V_m \cos(2\pi f_c t) \qquad (12.12)$$

which will be amplified to provide signal and reference levels

$$V_s = V_m A_s \cos(2\pi f_c t) \qquad (12.13)$$

and

$$V_R = V_m A_R \cos\left(2\pi f_c t + \frac{2\pi f_c D}{c}\right) \qquad (12.14)$$

where A_s and A_R are the voltage gains of the amplifiers used in the signal and reference paths between the mixer and the DBM.

For a loss-free DBM and low-pass filter, the output voltage that this produces will be

$$V_0 = V_m A_s \sin\left(\frac{2\pi \Delta f D}{c}\right). \qquad (12.15)$$

This can be increased using an amplifier whose voltage gain is A to obtain a change in the bias voltage

$$V_c = \frac{V_m A_s A 2\pi \Delta f D}{c} \qquad (12.16)$$

where it is assumed that $\Delta f \ll c/4D$.

Hence the actual change in oscillator frequency, when this change in bias is taken into account, will be

$$\Delta f = \frac{\delta f}{\delta V} V_c + \frac{\delta f}{\delta T} \Delta t. \qquad (12.17)$$

If we define the rate $\delta f/\delta V$ of change in frequency with bias voltage to be equivalent to the coefficient k, then we can re-express the above as

$$\Delta f = \frac{k V_m A_s A 2\pi \Delta f D}{c} + \Delta f(T) \qquad (12.18)$$

where $\Delta f(T)$ is the frequency change which would arise if the control system were absent. Now this arrangement may be regarded as a form of feedback loop. Hence we can say that

$$\Delta f(1 - G) = \Delta f(T) \qquad (12.19)$$

where G can be defined as a *loop gain*:

$$G \equiv \frac{k V_m A_s A 2\pi D}{c}. \qquad (12.20)$$

Looking at this result we can see that, if $|G| \gg 1$, then $\Delta f \ll \Delta f(T)$. By obtaining a system whose loop gain is sufficiently high we can expect to reduce any tendency for the oscillator frequency to alter. It should be noted that, whilst we can significantly reduce any frequency errors using this method, it is not possible to reduce Δf to zero for any finite value of G.

In practice it is generally important to ensure that G has the correct sign, i.e. that the change produced by the frequency control loop should have an opposite sign to that of the initial error. The wrong sign for G will tend to cause the frequency error to be *larger* than if the control loop were absent.

The analysis given above is satisfactory only when the size of the frequency change is small compared with $c/4D$ and we are justified in assuming that $\sin(2\pi \Delta f D/c) \simeq 2\pi \Delta f D/c$. For larger frequency changes we can see that $\sin(2\pi \Delta f D/c)$ will reach a maximum value when $\Delta f = 4c/D$. Any further increase in Δf will cause $\sin(2\pi \Delta f D/c)$ to fall, reaching zero when $\Delta f = 2c/D$.

Frequency Control Loops and Diplexers

As a consequence the system is non-linear in its response to sufficiently large frequency errors and the effective value of the loop gain, G, will tend to fall with increasing frequency changes. In particular, it should be noted that when $\Delta f > 2c/D$ the sign of G becomes inverted and the loop will tend to aggravate any frequency changes. Hence this type of system only operates as required over a limited frequency range.

A control loop of this general form is often called a *frequency lock loop* as it attempts to 'lock' the oscillator frequency f_0 into a particular relationship with the harmonics of the reference frequency f_R. The finite range of operation is referred to as the *lock-in range*. If, when the system is switched on, the initial frequency error is greater than this value, the loop will be unable to work correctly and to 'pull in' the oscillator frequency towards the required value.

In a real system, further problems may arise owing to the finite times taken for signals to move around the loop and the presence of a low-pass filter to average the output from the DBM. In practice, the oscillators, harmonic mixer, amplifiers, etc, will all generate some noise over a wide frequency range. Noise produced by the millimetre-wave oscillator will also appear as sidebands on either side of the mean oscillation frequency.

As a result, the voltages around the loop will have a noise spectrum superimposed upon them, some of which represent real fluctuations of the millimetre-wave oscillator frequency. At an appropriately high noise frequency these fluctuations will take a half-cycle or more to move around the loop. As a result, unless G is much less than unity at this frequency, there will be a tendency for the noise to be emphasized.

In some cases this may lead to the creation of 'peaked' sideband noise on the oscillator output or even—in serious cases—to an oscillator whose output frequency is modulated uncontrollably over a wide range. This behaviour is directly analogous to the onset of oscillations in an amplifier controlled by negative feedback. The loop has satisfied the Barkhausen criterion and may, itself, act as an unwanted oscillator. The solution is to alter the phase and time delays around the loop or to reduce the high-frequency value of the loop gain G to ensure that the criterion can no longer be met.

For any real loop, we shall obtain an optimum value for the gain G and the *loop bandwidth* B (the range of modulation frequencies within which $|G| > 1$). This means that we cannot expect the loop to deal with short-term frequency fluctuations which take place in a time scale less than $1/2B$ s.

Many real control loops take a form slightly different to that shown above. These replace the delayed signal input with a signal from another reference oscillator. (This may sometimes also be taken from a lower harmonic of the same oscillator used for the harmonic mixer.) The behaviour of such a system is broadly similar to that considered above.

Now, however, the frequency of the reference signal does not alter along with f_c.

In this alternative form of control loop the DBM compares the phase of f_c with that of the new reference oscillator output. The signal input presented to the DBM will now be

$$V_s = A_s V_m \cos(2\pi f_c t) \qquad (12.21)$$

and the reference input

$$V_R = B \cos(2\pi f_b t + \theta) \qquad (12.22)$$

where θ is the phase difference between the two inputs at the time $t = 0$.

The reference input frequency f_b is chosen to be equal to the value of f_c which emerges from the harmonic mixer when the LO frequency is f_0. The mean output voltage of the DBM will therefore be

$$V_0 = A_s V_m \cos(2\pi \Delta f t + \theta). \qquad (12.23)$$

As before, Δf represents the actual frequency error caused by some unwanted effect. Using the same example as before—a frequency change induced by temperature sensitivity—we can say that

$$\Delta f = kAA_s V_m \cos(2\pi \Delta f t + \theta) + \Delta f(T) \qquad (12.24)$$

$\Delta f(T)$ being the frequency change which would occur if the control loop were absent.

If we examine equation (12.24), it can be seen that any non-zero value of Δf must be time dependent. In order for the loop to work as required we must ensure that, when $\Delta f = 0$, $V_0 = 0$. This implies that $\theta = \pm \frac{1}{2}\pi$. Hence we can say that for a small frequency error

$$\Delta f = kAA_s V_m 2\pi \Delta f t + \Delta f(T) \qquad (12.25)$$

i.e.

$$\Delta f = \frac{\Delta f(T)}{1 - Gt} \qquad (12.26)$$

where

$$G = kAA_s V_m 2\pi. \qquad (12.27)$$

Hence, for a large negative value of G, we can say that, after a sufficiently large time,

$$\Delta f = \frac{\Delta f(T)}{Gt}. \qquad (12.28)$$

After a suitably long time, therefore, the actual frequency error Δf will have fallen virtually to zero.

This system acts by comparing the phase of the signal with that from a reference oscillator. As a consequence, this type of control system is generally referred to as a phase lock loop (PLL). Phase lock loops are often used in preference to a frequency lock loop as they can, in theory, provide a greater degree of control. This is because the control signal is obtained from a phase difference which may remain even when the frequencies become identical. This contrasts with the frequency lock loop system which always requires a non-zero frequency error Δf in order to produce a correcting voltage.

Phase lock loops tend to be more difficult to operate than frequency lock loops. In the description given above of how the loop operates it was assumed that Δf is small. If a frequency error arises which is too large then the output from the DBM will vary sinusoidally with the frequency Δf. For a large enough frequency error this variation will be suppressed by the low-pass filter at the output of the DBM and hence will not produce any corresponding control voltage. The system will be unable to 'lock' unless Δf is sufficiently small.

Comparing equations (12.19) and (12.26) we can see that the two expressions would be the same if, for the phase lock loop, we were to regard the effective gain as Gt instead of G. The effective gain of the phase lock loop can therefore be considered as if it were time dependent, tending to rise towards infinity as time passes. Although this comparison does not strictly compare like with like, it does serve to indicate that a phase lock loop acts in some ways as a frequency loop whose gain is very high. As a result, phase loops tend to be more prone to uncontrolled loop oscillations or sideband noise emphasis than frequency loops are.

A variety of frequency and phase lock systems have been evolved, each with its own detailed advantages and disadvantages. Some systems act as a hybrid of the phase and frequency loops, seeking to combine the best points of each. Each of these systems can be treated as a modification of the two basic arrangements described above.

Most practical systems employ the harmonic mixer technique. Whilst this is generally a good approach, it does suffer from two potential drawbacks: firstly, the use of a harmonic of the reference frequency f_R means that any uncertainty in our knowledge of f_R will be multiplied by the harmonic number to produce a large uncertainty in the relationship between f_c and f_0. Secondly, neither the sign of $f_0 - nf_R$ nor the correct value of the harmonic number are immediately apparent from the output. As a result, some care must be taken to ensure that we can determine unambiguously the value of f_0.

It is possible to avoid these difficulties by using frequency control systems which do not make use of a harmonic mixer. An example of such a system is illustrated in figure 12.4. The arrangement shown is based upon the use of a ring resonator, but any equivalent Fabry–Perot or similar resonator could also be employed to obtain the same effect.

In this system the signal levels transmitted and reflected by an optical resonator are measured by a pair of power detectors. The resonator is adjusted so as to have a reflectivity and transmissivity of 0.5 at the desired oscillation frequency f_0 and the detectors are 'matched', i.e. they are arranged to have identical sensitivities. The difference between the output voltages produced by the two detectors is then amplified to supply a sensed output level V.

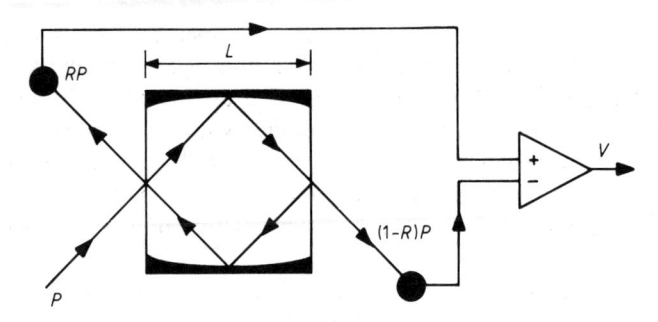

Figure 12.4 Ring resonator used as a frequency discriminator.

To see how this arrangement works we can take as a simple example a resonator whose power reflectivity R and transmissivity T vary as

$$R = \exp\left[-\left(\frac{f-f'}{\alpha}\right)^2\right] \qquad (12.29)$$

and

$$T = 1 - R \qquad (12.30)$$

i.e. the resonator has a resonant peak reflectivity at the frequency f', the variation in R with f has a Gaussian form, and the power reflectivity and transmissivity will be 0.5 at the frequencies $f = f' \pm \Delta f$, where

$$\Delta f = \alpha_{\backslash} - \ln 0.5. \qquad (12.31)$$

The quality factor Q for such a resonator will be $Q = f'/\Delta f$; hence the value of α will be such that

$$\alpha = 1.2011 \frac{f'}{Q}. \qquad (12.32)$$

Hence we can obtain α from knowing either Δf or f' and Q.

Having adjusted the system such that the desired oscillation frequency occurs at, say, $f_0 = f' - \Delta f$, then an input power level P will direct levels of $\frac{1}{2}P$ onto both the detectors. Any small change δf in the oscillator frequency will alter the reflected power P_R and transmitted power P_T, to

$$P_R = \tfrac{1}{2}P + \delta P \qquad P_T = \tfrac{1}{2}P - \delta P \qquad (12.33)$$

where

$$\delta P = \frac{2P\Delta f}{\alpha^2} \exp\left[-\left(\frac{\Delta f}{\alpha}\right)^2\right] \delta f. \qquad (12.34)$$

As a result we can expect that the output voltage V will be such that

$$V = \frac{4\mathcal{R}AP\Delta f}{\alpha^2} \exp\left[-\left(\frac{\Delta f}{\alpha}\right)^2\right] \delta f \qquad (12.35)$$

where $\mathcal{R}$ represents the sensitivity or responsivity of each detector and A represents the voltage gain of the differential amplifier. Knowing that $\exp[-(\Delta f/\alpha)^2] = 0.5$, the above expression can conveniently be rewritten as

$$V = \frac{2\mathcal{R}APQ}{f'(1.2011)^2} \delta f. \qquad (12.36)$$

The arrangement therefore provides an output voltage V whose magnitude and sign are essentially proportional to the amount δf by which the oscillator frequency departs from the required value f_0. Hence V can be used as a control voltage to adjust the operation of the oscillator and to correct its output frequency.

Whilst an optical arrangement of this sort is, in principle, quite simple compared with one based upon a harmonic mixer, it does suffer from some practical difficulties. The main problems arise when a high degree of frequency sensitivity is required in order to sense (and correct) very small frequency changes. This implies that a resonator with a large Q value is required.

Most millimetre-wave optical resonators have Q values up to the order of 100 or perhaps 1000. Whilst higher values, above 10 000, can be obtained, resonators of such high Q are difficult to build and couple into a system. As a result it is not easy to employ this technique to measure frequency changes which are significantly smaller than a few tens of megahertz for values of f_0 around 100 GHz.

Another alternative is to employ an MPI as a frequency discriminator. The path difference D of the interferometer is set to be equal to $(n \pm \frac{1}{4})\lambda_0$, where λ_0 is the free-space wavelength of the required oscillator frequency. For an input power P at the oscillator frequency f this will produce power levels

$$P_1 = P \sin^2\left(\frac{\pi D}{\lambda}\right) \qquad (12.37)$$

and

$$P_2 = P \cos^2\left(\frac{\pi D}{\lambda}\right) \qquad (12.38)$$

at its two output ports. These powers can then be coupled onto two similar detectors. For detectors of sensitivity $\mathcal{R}$ this will produce output voltages

$$V_1 = \tfrac{1}{2}\mathcal{R}P\left[1 - \cos\left(\frac{2\pi D}{\lambda}\right)\right] \quad (12.39)$$

and

$$V_2 = \tfrac{1}{2}\mathcal{R}P\left[1 + \cos\left(\frac{2\pi D}{\lambda}\right)\right] \quad (12.40)$$

when the oscillator frequency $f = c/\lambda$.

Using a differential amplifier of voltage gain A, we can obtain an output voltage

$$V = A(V_2 - V_1) = \mathcal{R}AP\cos\left(\frac{2\pi Df}{c}\right) \quad (12.41)$$

where $f_0 = c/\lambda_0$ is the nominal required oscillator frequency.

For a frequency error $\Delta f = f - f_0$, we can rewrite this as

$$V = \mathcal{R}AP\sin\left(\frac{2\pi D\Delta f}{c}\right) \quad (12.42)$$

which for a frequency error small enough to satisfy the inequality

$$\Delta f \ll \frac{c}{D} \quad (12.43)$$

is essentially equivalent to

$$V = \frac{\mathcal{R}AP2\pi D\Delta f}{c}. \quad (12.44)$$

This system is an almost perfect analogue of the use of a delay line and DBM, its sensitivity depending upon the delay line length D.

In practice it is quite straightforward to design and build MPI systems with path differences D of around 1 m or more. Hence, for frequencies in the few hundred gigahertz range it is possible to obtain n values of the order of 1000. Although not appropriate in all cases, an optical system such as the MPI is useful in a number of laboratory or other scientific measurement situations. This is because the system is simple in operation, offering a well defined level of performance over a frequency range from below 100 GHz to approaching 1 THz.

The level of frequency control which can be obtained using such a system is largely set by the signal-to-noise ratio of the detected signals. Given a ratio in excess of 100 to 1 for the maximum power level P and a value of $n \simeq 1000$, it is possible to detect and correct frequency changes of around 1 part in 100 000 (i.e. 1 MHz at 100 GHz). The absolute accuracy of frequencies set in this way will depend upon how well the actual path difference is known. Using standard laser ranging techniques an accuracy of

Frequency Control Loops and Diplexers

1 part in 1 000 000 or better can be obtained for a path difference around 1 m. These levels of frequency control and absolute accuracy are quite adequate for many purposes.

12.2 Gaussian Analysis of Fabry–Perot 'Ring' Diplexers

The function of a diplexer is to couple both LO and signal power efficiently onto the mixer of a heterodyne receiver system. Various forms of optical diplexer have been employed for millimetre-wave systems. One of the simplest consists of a thin sheet or slab of a suitable dielectric. By employing a piece of dielectric whose amplitude reflectivity is r we can, for example, arrange to couple LO power into a mixer via reflection by the dielectric and to couple signal power via transmission.

In most cases of practical interest the LO and signal frequencies will be similar. Hence we can generally expect that a thin sheet will have much the same reflectivity for both frequencies. The LO power-coupling efficiency N_l and signal power-coupling efficiency N_s will therefore be

$$N_l = |r|^2 \qquad N_s = 1 - |r|^2 \qquad (12.45)$$

(assuming that the system is otherwise without loss), i.e.

$$N_l + N_s = 1. \qquad (12.46)$$

As a consequence, no matter what value we choose for the reflectivity r, it appears that we cannot make either efficiency approach unity without reducing the other to almost zero!

In order to avoid this problem we must employ some form of diplexer whose properties are frequency dependent. It has already been mentioned in chapter 2 and chapter 8 that the reflectivity of a dielectric slab depends upon the relationship between the slab thickness and the radiation wavelength. One possible solution therefore would be to choose an effective slab thickness t such that

$$2t = \frac{(m \pm \tfrac{1}{2})\lambda_s}{\mu} = \frac{m\lambda_l}{\mu} \qquad (12.47)$$

where m is an integer, and λ_s and λ_l are the free-space wavelengths of the signal and LO inputs.

Taking a practical example, we can choose values of, say, $f_s = 250$ GHz ($\lambda_s = 1.199$ mm) for the signal and $f_l = 255$ GHz ($\lambda_l = 1.176$ mm) for the LO. Equation (12.47) can be rearranged to obtain

$$m = \pm \frac{\lambda_s}{2(\lambda_s - \lambda_l)} \qquad (12.48)$$

which, for the example values, implies that $m \simeq 25.58$. The closest integer value is therefore $m = 26$ for which we would require a slab thickness (assuming that $\mu \simeq 1.5$) of $t \simeq 20$ mm.

Whilst such a slab could be employed as a diplexer, its performance remains far from ideal. A slab of, for example, HDPE, which is 20 mm thick will have an absorptivity of 0.1 at 250 GHz. Hence we may expect such a slab to have a maximum transmissivity below 90% even when the thickness is chosen to give minimal reflection at this frequency. Furthermore, the peak power reflectivity of an HDPE slab is only around 16%. This combination of significant transmission loss with poor peak reflectivity means that a 'tuned' dielectric slab is not generally useful as a diplexer.

A number of millimetre-wave systems make use of one form or another of the Fabry–Perot resonator as a frequency-sensitive diplexer. 'Ring' resonators, like that used as part of a frequency control system earlier in this chapter, have a number of advantages over the basic form of Fabry–Perot resonator and have been widely used as diplexers.

Figure 12.5 illustrates the use of a ring resonator as a diplexer. To illustrate this we can assume that the resonator uses one inductive and one capacitive mesh, each having a signal reflectivity whose modulus is $|r|$. From chapter 8 we can say that the maximum and minimum reflectivity when illuminated with a single-mode beam will be

$$|\Gamma(\max)| = \frac{2|r|}{1+|r|^2} \qquad (12.49)$$

$$|\Gamma(\min)| = 0. \qquad (12.50)$$

The maximum reflectivity will occur for signals whose wavelength is such that the phase shift in transit between one mesh and the other is

$$\delta(\max) = n\pi + \tfrac{1}{2}\pi \qquad (12.51)$$

where n is an integer. The minimum reflectivity occurs for signals whose wavelength is such that the phase shift is

$$\delta(\min) = m\pi. \qquad (12.52)$$

We can now arrange the ring to have a maximum reflectivity at, say, the signal frequency, and a minimum reflectivity at the LO frequency. On the assumption that the input and output beams are single mode and have the appropriate beam waist size and location, this implies that the signal and LO coupling efficiencies will be

$$N_s = |\Gamma(\max)|^2 = \frac{4|r|^2}{(1+|r|^2)^2} \qquad (12.53)$$

$$N_l = |1 - \Gamma(\min)|^2 = 1. \qquad (12.54)$$

From equation (12.53) we can see that $|r|$ must be relatively high (i.e. approaching unity) if we require $N_s \simeq 1$.

Frequency Control Loops and Diplexers

When $|r|$ is nearly unity, we find that $|\Gamma|$ is also generally high, only falling towards zero over small ranges of signal wavelengths for which the phase shift in transit between meshes is close to $\delta(\min)$.

In practice the LO signal will essentially consist of a single frequency. However, the required signal will usually contain components spread over a finite range of frequencies or wavelengths. For this reason it is normally advisable to couple signal into the mixer by reflection when a resonant diplexer is employed. Otherwise, some parts of the signal spectrum may not couple efficiently through the diplexer.

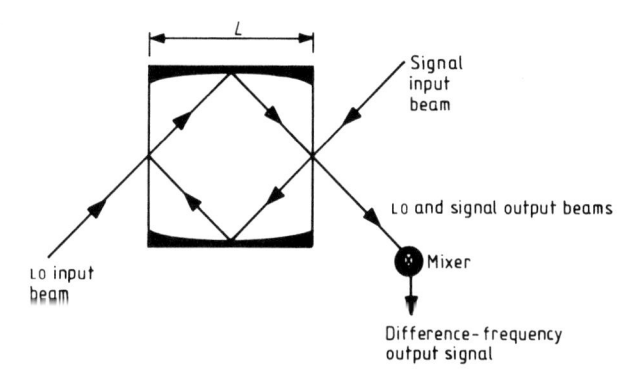

Figure 12.5 Ring resonator used as a diplexer.

Combining equations (12.51) and (12.52) we can define the required cube side size L for a suitable diplexer to be

$$L \simeq \frac{(p+\frac{1}{2})\lambda_i}{2\sqrt{2}} \tag{12.55}$$

where

$$\frac{1}{\lambda_i} = \left| \frac{1}{\lambda_s} - \frac{1}{\lambda_l} \right| \tag{12.56}$$

and p is an integer.

Here λ_l and λ_s are the LO and signal free-space wavelengths corresponding to the LO frequency f_l and signal frequency f_s, and it is assumed that $\lambda_s - \lambda_l$ is small enough to regard

$$\arctan\left(\frac{\lambda_s L}{\sqrt{2\pi\omega_0(f_s)^2}}\right) - \arctan\left(\frac{\lambda_l L}{\sqrt{2\pi\omega_0(f_l)^2}}\right) \simeq 0. \tag{12.57}$$

Taking the values previously chosen ($f_s = 250$ GHz; $f_l = 255$ GHz) as an example, we would require a ring in the form of a cube whose size would be

$$L \simeq (p+\tfrac{1}{2})21.6 \text{ mm}. \tag{12.58}$$

The smallest theoretical size of L (for $p = 0$) would then be around 10.5 mm. Given that the signal and LO wavelengths are both just over 1 mm this size is, however, probably too small to avoid problems with beam truncation and distortion. Hence a larger value of L (i.e. $p \geq 1$) would be preferable unless an exceptionally compact diplexer was required for some reason.

In fact, it is not always essential to choose a ring whose size satisfies both equation (12.51) and equation (12.52) for the signal and LO frequencies. This is because, when $|r|$ is high, the reflectivity remains high over wide ranges of frequencies centred upon the values which satisfy equation (12.51). Despite this, it is usually straightforward to find an acceptable cube size using the approach outlined above.

12.3 The Martin–Puplett Interferometer as a Diplexer

The *Martin–Puplett interferometer* (MPI) can also be used as an optical diplexer system. Although generally larger than a resonator, it has a number of advantages. Firstly, its performance can be analysed more simply than a resonator. As a consequence, the actual performance of a real instrument can be predicted with confidence and the design process is somewhat clearer. Secondly, the MPI operates inherently in a broad-band manner. As a result, a system designed for a particular use can usually be employed for a wide range of signal and LO frequencies simply by adjusting the location of one roof mirror. MPI diplexers designed for, say, 250 GHz, are often able to offer high levels of performance over the whole 100–500 GHz range without modification.

Figure 12.6 illustrates an MPI employed as a diplexer. In this example, the LO beam passes into the circuit via transmission through a 'horizontal' polarizer. The signal beam enters the circuit via reflection from this polarizer. In this way the two inputs are combined to produce a total input beam which is composed of two orthogonally polarized components. One component corresponds to the signal input, and the other corresponds to the LO. The total input beam field can therefore be represented by the vector

$$E = A_1 \exp(j2\pi f_1 t) V + A_s \exp(j2\pi f_s t) H \qquad (12.59)$$

where A_1 and f_1 represent the amplitude and frequency of the LO beam, and A_s and f_s represent the amplitude and frequency of the signal beam.

Coherent mixers respond to variations in the total applied field and in general only respond to that portion of the incident field parallel to a given axis. For a mixer which only responds to the field component whose electric vector is in the direction $(\cos \theta, \sin \theta)$ the apparent input field magnitude E_m

produced by E would be

$$E_m = A_l \cos\theta \exp(j2\pi f_l t) + A_s \sin\theta \exp(-j2\pi f_s t). \tag{12.60}$$

Clearly, in order to maximize the level of the difference-frequency output from the mixer we would wish to couple the available signal and LO into the mixer as efficiently as possible. However, equation (12.60) indicates that any choice of θ which maximizes one input will reduce the other to zero. In order to avoid this problem we must alter the polarization of one (or both) input components so that they share the same polarization state. We can then hope to couple all the available field into a mixer which responds to the chosen state.

The operation of the MPI as a diplexer can hence be visualized in terms of its action upon the polarization states of the signal and LO input beams. The beams enter the circuit along a common beam axis but are orthogonally polarized. They emerge along a common axis but should now have been altered so as to have a common polarization state.

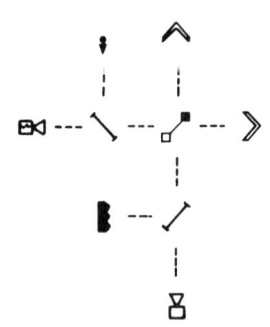

Figure 12.6 Martin–Puplett interferometer used as a diplexer.

In order to achieve this, the path difference must cause one beam to emerge with the same polarization state as it would have if the path difference were zero—but the other beam must emerge with its polarization orthogonal to that which it would have if the path difference were zero. The mixer can now be aligned to respond to plane polarized fields whose electric vectors are parallel to either V or H—whichever is appropriate.

Using the methods described in chapter 10 we can say that the total beam field directed onto the mixer or detector via the diplexer will be

$$O = ([Z_1]H_T P_T R P_R + [Z_2]H_T N_R R P_T)E. \tag{12.61}$$

This output beam will be plane polarized in the 'vertical' direction by its passage through the final 'horizontal' polarizer—any horizontal component will be reflected away from the mixer by this polarizer. The total output field directed onto a mixer can therefore be represented by the scalar value

$$O_v = \tfrac{1}{2} A_l \exp(j2\pi f_l t) \left[\exp\left(-\frac{j2\pi Z_2}{\lambda_l}\right) + \exp\left(-\frac{j2\pi Z_1}{\lambda_l}\right)\right]$$
$$+ \tfrac{1}{2} A_s \exp(j2\pi f_s t) \left[\exp\left(-\frac{j2\pi Z_2}{\lambda_s}\right) - \exp\left(-\frac{j2\pi Z_1}{\lambda_s}\right)\right].$$
(12.62)

This can conveniently be rewritten in the form

$$O_v = C_l A_l \exp(j2\pi f_l t) + C_s A_s \exp(j2\pi f_s t) \qquad (12.63)$$

where

$$C_l = \tfrac{1}{2} \left[\exp\left(-\frac{j2\pi Z_2}{\lambda_l}\right) + \exp\left(\frac{j2\pi Z_1}{\lambda_l}\right)\right] \qquad (12.64)$$

and

$$C_s = \tfrac{1}{2} \left[\exp\left(-\frac{j2\pi Z_2}{\lambda_s}\right) - \exp\left(-\frac{j2\pi Z_1}{\lambda_s}\right)\right] \qquad (12.65)$$

can be seen as amplitude coupling coefficients which indicate the efficiency of LO and signal coupling to the mixer. In order to maximize the signal and LO power coupling we must ensure that $|C_l|^2 \simeq 1$, and $|C_s|^2 \simeq 1$. These conditions may be combined to obtain

$$|C_l|^2 |C_s|^2 \simeq 1. \qquad (12.66)$$

Equation (12.66) can now be taken as the requirement which must be satisfied in order that the circuit should act as a diplexer, efficiently coupling the signal and LO powers simultaneously onto the mixer.

If we define the path difference $\Delta = Z_1 - Z_2$, then equations (12.64) and (12.65) can be used to obtain the expressions

$$|C_s|^2 = \tfrac{1}{4}\left\{\left[1 - \cos\left(\frac{2\pi\Delta}{\lambda_s}\right)\right]^2 + \sin^2\left(\frac{2\pi\Delta}{\lambda_s}\right)\right\} \qquad (12.67)$$

and

$$|C_l|^2 = \tfrac{1}{4}\left\{\left[1 + \cos\left(\frac{2\pi\Delta}{\lambda_l}\right)\right]^2 + \sin^2\left(\frac{2\pi\Delta}{\lambda_l}\right)\right\}. \qquad (12.68)$$

It may be seen from the form of these expressions that equation (12.66) will be satisfied provided that

$$\cos\left(\frac{2\pi\Delta}{\lambda_l}\right) \simeq 1 \qquad \cos\left(\frac{2\pi\Delta}{\lambda_s}\right) \simeq -1 \qquad (12.69)$$

Frequency Control Loops and Diplexers

i.e. the above condition can be simplified into the form

$$\cos\left(\frac{2\pi\Delta}{\lambda_l}\right)\cos\left(\frac{2\pi\Delta}{\lambda_s}\right) \simeq -1. \quad (12.70)$$

Using standard trigonometric identities this can be changed to

$$\cos\left[2\pi\Delta\left(\frac{1}{\lambda_l}-\frac{1}{\lambda_s}\right)\right] + \cos\left[2\pi\Delta\left(\frac{1}{\lambda_l}+\frac{1}{\lambda_s}\right)\right] \simeq -2. \quad (12.71)$$

In most cases of practical interest, $|1/\lambda_l - 1/\lambda_s| \ll |1/\lambda_l + 1/\lambda_s|$, and the second cosine term varies periodically with Δ much more swiftly than the first term does. As a consequence we may expect that any path difference value which satisfies the requirement

$$\cos\left[2\pi\Delta\left(\frac{1}{\lambda_l}-\frac{1}{\lambda_s}\right)\right] \simeq -1 \quad (12.72)$$

will be very close to a value which satisfies equation (12.71). Now the difference (or intermediate) frequency $f_i = |f_s - f_l|$ will have a corresponding free-space wavelength λ_i such that

$$\frac{1}{\lambda_i} = \left|\frac{1}{\lambda_l} - \frac{1}{\lambda_s}\right|. \quad (12.73)$$

Combining this with the previous expression we find that the required path difference will be

$$\Delta \simeq \tfrac{1}{2}\lambda_i. \quad (12.74)$$

The actual operating value of Δ can then be found by slightly adjusting the path difference about this value until the second cosine term in equation (12.71) is also approximately equal to -1, i.e. until the signal and LO powers reaching the mixer have been maximized.

It is interesting to note that the value of Δ does not depend very strongly upon f_s or f_l but is mainly dependent upon the difference frequency $|f_s - f_l|$. As a consequence, provided that the difference frequency remains unchanged, this form of diplexer can be used for a wide range of signal frequencies with only a slight adjustment to the path difference setting.

12.4 Gaussian Analysis of the Martin–Puplett Interferometer as a Diplexer

The analysis given above is satisfactory to understand the basis of the circuit's operation. In order to predict the detailed performance of a specific compact system we can make use of Gaussian beam mode techniques.

Using the same arguments as in chapter 9 we can say that an input signal beam of field ψ_s and LO beam of field ψ_l will produce field components

$$\Psi_s = \tfrac{1}{2}[\psi_s(Z) + \mathcal{T}\psi_s(Z + \Delta)] \tag{12.75}$$
$$\Psi_l = \tfrac{1}{2}[\psi_l(Z) - \mathcal{T}\psi_l(Z + \Delta)] \tag{12.76}$$

at the mixer output port. Here Z is taken to represent the distance between the beam waist locations and the chosen output port plane and Δ is the path difference. The value of $\mathcal{T}$ will be $+1$ or -1 depending upon which output port is coupled to the mixer.

For example we can consider the case where both the signal and the LO input beams are composed only of a circularly symmetric fundamental Gaussian mode and that both beams have the same beam waist ω_0. In this case, the signal and LO coupling efficiencies between the input and output ports will be

$$N_s = \tfrac{1}{2}[1 + \mathcal{T}\,\mathrm{Re}(\beta_s)] \tag{12.77}$$

and

$$N_l = \tfrac{1}{2}[1 - \mathcal{T}\,\mathrm{Re}(\beta_l)] \tag{12.78}$$

where

$$\beta_s = \frac{\exp(j2\pi\,\Delta/\lambda_s)}{1 + j\Delta\lambda_s/2\pi\omega_0^2} \tag{12.79}$$

and

$$\beta_l = \frac{\exp(j2\pi\,\Delta/\lambda_l)}{1 + j\Delta\lambda_l/2\pi\omega_0^2}. \tag{12.80}$$

In most cases, $\lambda_s \simeq \lambda_l$; hence the maximum and minimum values that we can obtain for N_s and N_l are almost identical.

Using the same example as in the section on ring diplexers where $f_s = 250$ GHz and $f_l = 255$ GHz, we shall require a path difference Δ of about 30 mm. Both the signal and the LO wavelengths will be approximately 1.2 mm. Under these conditions a beam waist size of around 6 mm (five signal wavelengths) will provide a maximum power-coupling efficiency of 0.994 (i.e. an insertion loss of about -0.027 dB). A beam waist size of just three signal wavelengths (5.6 mm) will provide a maximum power-coupling efficiency of 0.957 (-0.190 dB).

These losses arise because the MPI splits each input field into two orthogonal components which are recombined after having travelled along paths of differing lengths. The effective beam sizes $\omega(Z)$ and $\omega(Z + \Delta)$ of the two components will differ at any subsequent location along the beam, i.e. their field distributions cannot be identical. As a consequence the total beam cannot be plane polarized in a single orientation. Some portion of the total field will be plane polarized 'cross polar' to the required orientation.

Frequency Control Loops and Diplexers

The power 'lost' in the above coupling calculation is not absorbed within the diplexer. It emerges from the circuit as a field component which is orthogonally polarized to that required for the mixer. Hence, in the system shown, it is reflected by the final polarizer and does not reach the mixer.

At first examination the final polarizer does not serve any useful function. Even if it were removed, the mixer would itself reject the cross polar field component. However, this rejected power would then largely be retro-reflected by the mixer (and its antenna) back towards the input ports. This is undesirable as reflected power may affect either the signal or the LO source. The final polarizer is therefore useful in ensuring that the lost power can be absorbed by a load placed at the output port which is not used for the mixer.

In practice, most LO sources will produce unwanted sideband noise over a wide range of frequencies centred upon the desired output frequency. If noise at the frequencies $f = f_l \pm f_i$ reaches the mixer, it will produce a corresponding level of noise in the mixer output at the required difference frequency f_i. Hence it is important to ensure that the level of LO sideband noise falling upon the mixer is small compared with the signal. Otherwise this noise will raise the system noise level and swamp the required signal output.

In a typical heterodyne system the required LO power level at the frequency f_l will be a few milliwatts. The input signal power level may, however, be much smaller. If we take an extreme example from radio-astronomy, a source whose effective brightness temperature is 10 K will produce a signal power of 1.4×10^{-16} W MHz^{-1} bandwidth. Hence the unwanted sideband noise level from the LO must be at least 10^{13} times smaller than the required output to avoid swamping the signal. This is a very demanding specification for a real oscillator and the level of LO sideband noise reaching the mixer may be the dominating contribution to the system noise level.

Now the MPI circuit is a symmetric reciprocal system. Any noise power from the LO port at the frequencies $f_l \pm f_i$ is initially orthogonally polarized to input signal power at these frequencies. It follows that this noise power will therefore also emerge orthogonally polarized to the signal. Hence this noise power will be reflected by the output polarizer and will not reach the mixer. As a consequence the diplexer serves the useful function of diverting LO noise at the signal and image frequencies away from the mixer.

This property of the diplexer to reject unwanted LO noise is very useful in practical systems. It can sometimes produce a dramatic reduction in the system noise level, making the receiver system considerably more sensitive. Just as we have defined the power-coupling efficiencies N_s and N_l for signal and LO coupling, we can also define a *rejection ratio R* which indicates the level of noise rejection the diplexer produces.

By symmetry, we can say that for a correctly adjusted MPI diplexer the power rejection ratio value will be

$$R = 1 - N_s \qquad (12.81)$$

for noise at the frequencies $f_l \pm f_i$.

For a system designed for use with a signal frequency of 250 GHz and LO frequency of 255 GHz we have already seen that the required path difference will be around 30 mm. A beam waist size of 6 mm then can provide a signal coupling loss of -0.027 dB ($N_s = 0.994$) which is negligible for most purposes. This means that the peak rejection ratio of the diplexer will be about 0.006 or -22 dB.

In general it would not be worth increasing the beam size simply to recover part of the 0.6% of signal power which is lost. However, increasing the beam size may significantly improve the rejection ratio. For example, increasing ω_0 to 12 mm in the above situation would change the rejection ratio to 34 dB, reducing the LO produced noise level at the mixer by an extra factor of more than 10. As a consequence, the size of a practical diplexer is often chosen in order to achieve a desired level of LO noise rejection and not simply to obtain a specific signal or LO coupling efficiency.

A real wire grid polarizer will typically have a cross polar leakage of around -30 to -40 dB depending upon how it was made and the signal frequency. Hence it is generally sensible to choose a beam waist size which offers a comparable rejection ratio. The sizes of the optical elements can then be chosen to obtain a negligible level of truncation losses by using the approach outlined in chapter 3.

12.4.1 The effects of finite signal bandwidth

In the arguments presented so far, it has been implicitly assumed that the required signal is essentially monochromatic. In practice the incoming signal is normally distributed over a particular bandwidth B centred at a nominal centre frequency f_s. If the diplexer has been set up to provide optimal coupling efficiency at f_s, we now find that the coupling efficiency for spectral components near the ends of this band is significantly lower than at f_s.

The spectral properties of a diplexer can be considered by observing how the power-coupling efficiency $N(f)$ varies with frequency, once $N(f_s)$ has been maximized.

For a resonant form of diplexer such as the ring resonator, $N(f)$ will be equivalent to the variation with frequency in the appropriate power transmissivity or reflectivity. The form of this variation will depend upon a number of factors. In particular, the coupling efficiency variations will depend upon the reflectivities of the semi-reflectors used to make the resonator. This can be an advantage in some circumstances as it becomes possible to alter the frequency dependence of the coupling efficiencies for specific purposes. The corresponding disadvantage is that the behaviour of a compact resonator can sometimes only be predicted following an involved calculation.

In contrast, the behaviour of an MPI diplexer can simply be predicted using equations (12.77)–(12.80). Here $N(f)$ always varies essentially sinusoidally with f. If the peak coupling efficiency at the centre frequency f_s is $N(f_s)$, then from the above expressions we can say that for a frequency $f = f + \delta f$ the coupling efficiency can be obtained using

$$N(f) \simeq N(f_s)\tfrac{1}{2}\left[1 + \cos\left(\pi \frac{\delta f}{f_i}\right)\right]. \tag{12.82}$$

If we take as an example a signal which extends over a bandwidth of 100 MHz and use a system where $f_i = 5$ GHz, we can calculate that for frequencies at the extreme end of the signal band ($\delta f = 50$ MHz) the power-coupling efficiency will be

$$N(f) \simeq N(f_s)0.9997. \tag{12.83}$$

Hence the change in coupling efficiency over the signal frequency bandwidth can be seen to be minimal provided that the bandwidth is reasonably small compared with the chosen intermediate frequency f_i.

The corresponding change in the LO noise rejection ratio for frequencies away from the nominal signal frequency can be obtained from

$$R(f) \simeq 1 - N(f). \tag{12.84}$$

For the system considered above and taking a beam waist size of 12 mm at a signal centre frequency of 250 GHz (i.e. $N(f_s) = 0.9996$) this implies that the rejection ratio for noise at frequencies near the ends of a 100 MHz signal bandwidth would be around -31 dB. This compares with -34 dB at the centre frequency.

Appendix 1

Beam Normalization and Mode Polynomials

The mode expressions quoted throughout this book are given in terms of either Hermite (for Cartesian coordinate systems) or generalized Laguerre (for circular coordinates) polynomials. The definitions of the polynomials employed are arranged such that each mode is of a form normalized to give unity integrated beam power. The required form of these polynomials is derived in this appendix.

A1.1 Cartesian Coordinates and Hermite Polynomials

We require the expression used for the mnth mode to be normalized such that

$$\langle \psi_{mn} | \psi_{mn} \rangle = 1 \tag{A1.1}$$

where

$$\psi_{mn} = \psi_m(x)\psi_n(y)\exp[-j(kz - 2\pi ft)] \tag{A1.2}$$

and

$$\psi_m(x) = \sqrt{\frac{\omega_{0x}}{\omega_x}} \cdot \mathcal{H}_m\left(x\frac{\sqrt{2}}{\omega_x}\right) \exp\left\{-j\Phi_m - x^2\left[\left(\frac{1}{\omega_x}\right)^2 + \frac{jk}{2R_x}\right]\right\} \tag{A1.3}$$

with a similar expression for $\psi_n(y)$.

The Hermite polynomials must therefore be such that

$$\langle \psi_m(x) | \psi_m(x) \rangle = \langle \psi_n(y) | \psi_n(y) \rangle = 1 \tag{A1.4}$$

i.e. we can require

$$\int_{-\infty}^{+\infty} \mathcal{H}_m^2\left(x\frac{\sqrt{2}}{\omega_x}\right) \exp\left(-\frac{2x^2}{\omega_x^2}\right) dx = 1. \tag{A1.5}$$

By consulting an appropriate book on integrals and orthogonal polynomials we may find the standard result

$$\int_{-\infty}^{+\infty} \mathcal{H}_m^2(t) \exp(-t^2) \, dt = 2^m m! \sqrt{\pi}. \tag{A1.6}$$

If we therefore make the substitution

$$t^2 = \frac{2x^2}{\omega_x^2} \tag{A1.7}$$

into equation (A1.6), we produce the result

$$\int_{-\infty}^{+\infty} \mathcal{H}_m^2(t) \exp(-t^2) \frac{\sqrt{2}}{\omega_x} \, dt = 2^m m! \sqrt{\pi} \tag{A1.8}$$

which clearly contradicts equation (A1.5). This problem arises because the standard form of Hermite polynomials is not normalized in such a way as to provide immediately the result that we require.

In order to resolve this contradiction and to obtain a correctly normalized expression for ψ_m, we could choose to redefine ψ_m in the form

$$\psi_m(x) = \alpha_m \mathcal{H}_m\left(x \frac{\sqrt{2}}{\omega_x}\right) \exp\left\{-j\Phi_m - x^2 \left[\left(\frac{1}{\omega_x}\right)^2 + \frac{jk}{2R_x}\right]\right\} \tag{A1.9}$$

where

$$\alpha_m = \left(2^m m! \omega_x \sqrt{\frac{\pi}{2}}\right)^{-1/2} \tag{A1.10}$$

A similar result may be obtained for $\psi_n(y)$.

Whilst this approach is frequently used, it introduces the need to keep writing extra factors into mode expressions simply in order to ensure correct

Table A1.1 α_m and a_p values for the first few Hermite polynomials.

m	\multicolumn{9}{c}{a_p for the following p values}	α_m
	0	1	2	3	4	5	6	7	8	
0	1									0.893 24
1		1								0.631 62
2	−2		4							0.315 81
3		−12		8						0.128 92
4	12		−48		16					0.045 58
5		120		−160		32				0.014 41
6	−120		720		−480		64			0.004 16
7		−1680		3360		−1344		128		0.001 11
8	1680		−13 440		13 440		−3584		256	0.000 27
etc										

Appendix 1

normalization. To avoid this an alternative approach is adopted in this book.

Here we choose to redefine the Hermite polynomials to include already the factor α_m but with the factor $\sqrt{1/\omega_x}$ moved outside into the expression. We can now define the mth-order Hermite polynomials that we require to be of the form

$$\mathcal{H}_m(t) = \alpha_m(a_0 + a_1 t + a_2 t^2 + \cdots + a_m t^m) \tag{A1.11}$$

where t is either $x\sqrt{2}/\omega_x$ or $y\sqrt{2}/\omega_y$ as required.

The appropriate α_m and a_p values for the first few Hermite polynomials are given in Table A1.1.

A1.2 Circular Coordinates and Laguerre Polynomials

For beams described in a cylindrical coordinate system the beam modes take the form of a Gaussian multiplied by a generalized Laguerre polynomial. We may make use of an argument similar to that employed for defining the Hermite polynomials for Cartesian coordinates.

As before, we define the pth mode ψ_m to have unity total beam power, i.e. we require

$$\langle \psi_{pl} | \psi_{pl} \rangle = 1. \tag{A1.12}$$

Here the mode number p determines the radial variation of the field and l determines how the field may change with rotation angle about the beam axis.

For simplicity we shall here deal only with modes which have circular symmetry, i.e. those for which $l = 0$. The pth mode of this form may be obtained from the expression

$$\psi_p = \frac{1}{\omega} L_p\left(\frac{2r^2}{\omega^2}\right) \exp[-j(kz - 2\pi ft)] \exp\left\{j\Phi_p - r^2\left[\left(\frac{1}{\omega}\right)^2 + \frac{jk}{2R}\right]\right\} \tag{A1.13}$$

where

$$\Phi_p = (2p + 1) \arctan\left(\frac{\lambda z}{\pi \omega_0^2}\right) \tag{A1.14}$$

and $L_p(t)$ is defined to be

$$L_p(t) = \beta_p(b_0 + b_1 t + b_2 t^2 + \cdots + b_p t^p). \tag{A1.15}$$

By reference to an appropriate book on integrals and orthogonal polynomials we may say that

$$\beta_p = \frac{\sqrt{2/\pi}}{p!}.$$

Table A1.2 gives b_m values for the first few modes.
Further values for b_m can be obtained using

$$b_m = \frac{(p!)^2}{(m!)^2(p-m)!}.$$

Table A1.2 b_m values for the first few modes.

p	\multicolumn{8}{c}{b_m for the following m values}							
	0	1	2	3	4	5	6	7
0	1							
1	1	−1						
2	2	−4	1					
3	6	−18	9	−1				
4	24	−96	72	−16	1			
5	120	−600	600	−200	25	−1		
6	720	−4 320	5 400	−2 400	450	−36	1	
7	5 040	−35 280	52 920	−29 400	7 350	−882	49	−1
etc								

Appendix 2

Circuit Symbols and Matrices

A2.1 Circuit Symbols

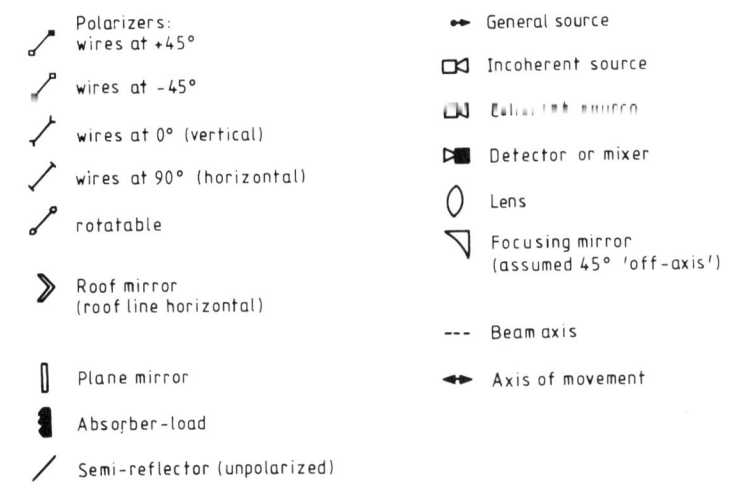

- Polarizers:
  wires at +45°
- wires at −45°
- wires at 0° (vertical)
- wires at 90° (horizontal)
- rotatable
- Roof mirror (roof line horizontal)
- Plane mirror
- Absorber-load
- Semi-reflector (unpolarized)

- General source
- Incoherent source
- Coherent source
- Detector or mixer
- Lens
- Focusing mirror (assumed 45° 'off-axis')
- Beam axis
- Axis of movement

Each of the above is shown in a form assuming that the input signal is moving from left to right. Note that in general it is assumed that coherent sources emit a signal whose electric field is plane polarized in the V direction and that detectors receive any signal coupled onto them.

A2.2 Jones Matrices

Input and output signals can be represented by a polarization vector of the form

$$E = (E_V, E_H) = E_V V + E_H H.$$

The effect of a polarizer whose wires appear at an angle θ to the vertical direction V when viewed along the beam line can then be described by a pair of matrices,

$$\begin{pmatrix} -\cos^2\theta & -\cos\theta\sin\theta \\ \cos\theta\sin\theta & \sin^2\theta \end{pmatrix} = \mathbf{A_R}$$

$$\begin{pmatrix} \sin^2\theta & -\cos\theta\sin\theta \\ -\cos\theta\sin\theta & \cos^2\theta \end{pmatrix} = \mathbf{A_T}$$

where $\mathbf{A_R}$ describes the effect of reflection and $\mathbf{A_T}$ describes the effect of transmission.

The properties of the four fixed polarizers corresponding to $\theta = 0°$, $90°$, $-45°$ and $+45°$ can now be defined in terms of the eight matrices given in Table A2.1.

Table A2.1 Properties of four fixed polarizers

Polarizer	Transmission	Reflection
Wires vertical, $\theta = 0°$	$\mathbf{V_T}$ $\begin{pmatrix} 0 & 0 \\ 0 & 1 \end{pmatrix}$	$\mathbf{V_R}$ $\begin{pmatrix} -1 & 0 \\ 0 & 0 \end{pmatrix}$
Wires horizontal, $\theta = 90°$	$\mathbf{H_T}$ $\begin{pmatrix} 1 & 0 \\ 0 & 0 \end{pmatrix}$	$\mathbf{H_R}$ $\begin{pmatrix} 0 & 0 \\ 0 & 1 \end{pmatrix}$
Wires at $\theta = +45°$	$\mathbf{P_T}$ $\frac{1}{2}\begin{pmatrix} 1 & -1 \\ -1 & 1 \end{pmatrix}$	$\mathbf{P_R}$ $\frac{1}{2}\begin{pmatrix} -1 & -1 \\ 1 & 1 \end{pmatrix}$
Wires at $\theta = -45°$	$\mathbf{N_T}$ $\frac{1}{2}\begin{pmatrix} 1 & 1 \\ 1 & 1 \end{pmatrix}$	$\mathbf{N_R}$ $\frac{1}{2}\begin{pmatrix} -1 & 1 \\ -1 & 1 \end{pmatrix}$

The effect of travelling a distance d can be represented by the matrix

$$\exp\left(-\frac{2\pi j d}{\lambda}\right)\begin{pmatrix} 1 & 0 \\ 0 & 1 \end{pmatrix} = [\mathbf{d}].$$

A roof mirror which is placed with its roof line horizontal can be represented by the reflection matrix

$$\begin{pmatrix} -1 & 0 \\ 0 & -1 \end{pmatrix} = \mathbf{R}$$

Appendix 2

and a plane mirror by the reflection matrix

$$\begin{pmatrix} -1 & 0 \\ 0 & 1 \end{pmatrix} = \mathbf{M}.$$

Although the matrices given above are sufficient for most purposes, it is also useful to define the matrices appropriate to a few additional devices. The most commonly useful additions are as follows.

(i) A roof mirror whose roof line is at an angle θ to the vertical can be represented by

$$\begin{pmatrix} \cos(2\theta) & -\sin(2\theta) \\ \sin(2\theta) & \cos(2\theta) \end{pmatrix} = \mathbf{R}(\theta).$$

By far the most common alternative to a roof mirror whose roof line is horizontal ($\theta = 90°$) is one with a vertical roof line ($\theta = 0°$). For convenience, $\mathbf{R}$ may be used to represent a roof mirror with $\theta = 90°$ in circuits where no other type of roof mirror is present. Mirrors with vertical and horizontal roof lines may be distinguished either by writing $\mathbf{R}(0°)$ and $\mathbf{R}(90°)$ or by using the notation $\mathbf{R_V}$ and $\mathbf{R_H}$. However, whilst this use of subscripts to indicate vertical or horizontal may be convenient, some care should be taken to avoid confusion with the subscripts R and T used to indicate reflection or transmission.

(ii) A general semi-reflector may be represented by the two matrices

$$\begin{pmatrix} r00 & r01 \\ r10 & r11 \end{pmatrix} = \mathbf{S_R}$$

to describe the effects of reflection, and

$$\begin{pmatrix} t00 & t01 \\ t10 & t11 \end{pmatrix} = \mathbf{S_T}$$

to describe the effects of transmission. For a semi-reflector whose properties are polarization insensitive, we can say that $r10 = r01 = t10 = t01 = 0$, $r00 = -r11$ and $t00 = t11$. The effects of absorption losses in a semi-reflector will cause $|r|^2 + |t|^2 < 1$.

Further Reading and Selected References

The following is a selection of some useful references. It is not intended as a comprehensive survey as it would be impossible to keep such a survey up to date. Information regarding recent work on millimetre-wave subjects often appears in the *IEEE Transactions on Antennas and Propagation, IEEE Transactions on Microwave Theory and Techniques* and *IEEE Transactions on Electron Devices* and in *Infrared Physics* or optical journals.

In recent years, the most consistently useful sources of information on millimetre-wave subjects have been published under the title *Infrared and Millimetre Waves*. This consists of a monthly journal and a series of case-bound volumes, all edited by K J Button and published by Academic Press. The journal provides reports of new work; the case-bound volumes contain review articles on specific subjects. These contain so many useful references that I have excluded any specific mentions from the following list as this would virtually require a complete index of these publications! Instead, the reader is urged to examine the contents of these volumes as a whole.

Books

Abramowitz M and Stegun I A 1965 *Handbook of Mathematical Functions* (New York: Dover)
Arnaud J A 1976 *Beam and Fiber Optics* (London: Academic)
Kraus J D 1988 *Antennas* (London: McGraw-Hill)
Schelkunoff S A 1943 *Electromagnetic Waves* (New York: Van Nostrand)

Millimetre-wave Optics, Devices and Systems

Papers

Antennas

Cummings H J and Frayne P G 1982 A photolithographic technique for reducing sub-mm antennas *Precis. Eng.* **4** 163–7

Dragone C 1977 Characteristics of a broadband microwave corrugated feed *Bell Syst. Tech. J.* **56** 869–88

Klein B J and Degnan J J 1974 Optical antenna gain *Appl. Opt.* **13** 2134–44

Silver S 1962 Microwave aperture antennas and diffraction theory *J. Opt. Soc. Am.* **52** 131–41

Thomas B MacA 1978 Design of corrugated conical horns *IEEE Trans. Antennas Propag.* **AP-26** 367–72

Thungren T, Kollberg E L and Yngvesson K S 1982 Vivaldi antennas for single beam integrated receivers *Proc. 12th Eur. Microwave Conf., Helsinki, 1982* 361–6

Wylde R J 1984 Millimetre-wave Gaussian beam mode optics and corrugated feed horns *IEE Proc.* H **131** 258–62

Resonant filters and diplexers

Anderson I 1975 On the theory of self-resonant grids *Bell Syst. Tech. J.* **54** 1725–31

Arnaud J A and Pelow F A 1975 Resonant grid quasi-optical diplexers *Bell Syst. Tech. J.* **54** 263–83

Arnaud J A, Saleh A A M and Ruscio J T 1974 Walk-off effects in Fabry–Perot diplexers *IEEE Trans. Microwave Theory Tech.* **MTT-22** 486–93

Baker E A M and Walker B 1982 Fabry–Perot interferometers for use at sub-mm wavelengths *J. Phys. E: Sci. Instrum.* **15** 25–32

Clark R N and Rosenberg C B 1982 Fabry–Perot and open resonators at microwave and mm-wave frequencies *J. Phys. E: Sci. Instrum.* **15** 9–24

Ulrich R 1967 Far-infrared properties of metallic mesh and its complemetary structure *Infrared Phys.* **7** 37–55

Wannier P G, Arnaud J A, Pelow F A and Saleh A A 1976 Quasi-optical band-rejection filter at 100 GHz *Rev. Sci. Instrum.* **47** 56–8

Gaussian beam modes

Arnaud J A 1971 Mode coupling in first-order optics *J. Opt. Soc. Am.* **61** 751–8

Arnaud J A and Kogelnik H 1969 Gaussian light beams with general astigmatism *Appl. Opt.* **8** 1687–93

Boyd G D and Gordon J P 1961 Confocal multimode resonator for millimetre through optical wavelength masers *Bell Syst. Tech. J.* **40** 489–508

Further Reading and Selected References

Dickson L D 1970 Characteristics of a propagating Gaussian beam *Appl. Opt.* **9** 1854–61

Feiock F D 1978 Wave propagation in optical systems with large apertures *J. Opt. Soc. Am.* **68** 485–9

Goubau G and Schwering F 1961 On the guided propagation of electromagnetic wave beams *IRE Trans. Antennas Propag.* **AP-9** 248–55

Kogelnik H 1964 Coupling and conversion coefficients for optical modes *Proc. Symp. Quasi-Optics, Brooklyn, 1964* (Brooklyn: Polytechnic Press) pp 333–47

Kogelnik H and Li T 1966 Laser beams and resonators *Proc. IEEE* **54** 1312–29

Martin D H and Lesurf J C G 1978 Sub-mm wave optics *Infrared Phys.* **18** 405–12

Williams C S 1973 Gaussian beam formulas from diffraction theory *Appl. Opt.* **12** 872–6

Two-beam interferometers and diplexers

Ade P A R, Costley A E, Cunningham C T, Mok C L, Neil G F and Parker T J 1979 Free-standing grids wound from 5 μm diameter wire for spectroscopy at far-infrared wavelengths *Infrared Phys.* **19** 599–601

Costley A E, Hursey K H, Neil G F and Ward J W M 1976 Free-standing wire grids *Proc. 2nd Conf. Sub-millimetre Waves and their Applications, Puerto Rico, 1976*

Erickson N R 1977 A directional filter diplexer using optical techniques for millimetre to submillimetre wavelengths *IEEE Trans. Microwave Theory Tech.* **MTT-25** 865–6

Lesurf J C G 1981 Gaussian optics and the design of Martin–Puplett diplexers *Infrared Phys.* **21** 383–90

—— 1988 Gaussian beam mode optics and the design of MPI instruments *Infrared Phys.* **28** 129–37

Martin D H and Puplett E 1969 Polarised interferometric spectrometry for the millimetre and submillimetre spectrum *Infrared Phys.* **10** 105–9

Payne J M and Wordeman M R 1978 Quasi-optical diplexer for millimetre wavelengths *Rev. Sci. Instrum.* **49** 1741–3

Peck E R 1962 Polarisation properties of corner reflectors and cavities *J. Opt. Soc. Am.* **52** 253–7

Lenses and dielectric materials

Alvarez J A, Jennings R E and Moorwood A F M 1975 Far-infrared measurements of selected optical materials at 1.6 °K *Infrared Phys.* **15** 45–9

Birch J R, Dromey J D and Lesurf J C G 1981 The optical constants of some common low-loss polymers between 4 and 40 cm^{-1} *Infrared Phys.* **21** 225–8

Birch J R and Lesurf J C G 1987 The near millimetre wavelength optical constants of Flurosint *Infrared Phys.* **27** 423–4

Chantry G W, Evans H M, Fleming J W and Gebbie H A 1969 TPX: a new material for optical components in the far infrared spectral region *Infrared Phys.* **9** 31–3

Jones E M T and Cohn S B 1955 Surface matching of dielectric lenses *J. Appl. Phys.* **26** 452–7

Mixers and detectors

Arams F, Allen C, Peyton B and Sard E 1966 Millimetre mixing and detecting in bulk InSb *Proc. IEEE* **54** 612–22

Archer J W and Mattauch R J 1981 Low-noise single ended mixer for 230 GHz *Electron. Lett.* **17** 180–1

Carlson E R, Schneider M V and McMaster T F 1978 Subharmonically pumped millimetre wave mixers *IEEE Trans. Microwave Theory Tech.* **MTT-26** 706–15

Jones R C 1953 The general theory of bolometer performance *J. Opt. Soc. Am.* **43** 1–14

Kerr A R 1975 Low-noise room-temperature and cryogenic mixers for 80–120 GHz *IEEE Trans. Microwave Theory Tech.* **MTT-23** 781–7

Larrabee R D, Woodard D W and Higinbothem W A 1965 Microwave impedance of semiconductor posts in waveguides. Parts I & II *J. Appl. Phys.* **36** 1597–663

Whalen J J and Westgate C R 1970 InSb Mixers *Proc. Symp. Millimetre Waves* (New York: Polytechnic) pp 305–20

Wrixon G T and Kelly W M 1978 Low-noise Schottky barrier diode mixers for wavelengths below 1 mm *Infrared Phys.* **18** 413–28

Oscillators

Brown E R, Goodhue W D and Sollner T C L G 1988 Fundamental oscillations up to 200 GHz in resonant tunneling diodes and new estimates of their maximum oscillation frequency from stationary-state tunneling theory *J. Appl. Phys.* **64** 1519–29

Brown E R, Sollner T C L G, Goodhue W D and Parker C D 1987 Millimetre-band oscillations based upon resonant tunneling in a double-barrier diode at room temperature *Appl. Phys. Lett.* **50** 83–5

Dunn D A 1958 Understanding the backward wave oscillator *Electron. Indust.* January, 73–6

Friscourt M-R, Rolland P A, Cappy A, Constant E and Salmer G 1983 Theoretical contribution to the design of millimetre wave TEO's *IEEE Trans. Electron. Devices* **ED-30** 223–9

Haddad G I, Greiling P T and Schroeder W E 1970 Basic principles and properties of avalanche transit-time devices *IEEE Trans. Microwave Theory Tech.* **MTT-18** 752–72

Further Reading and Selected References

Haydl W H 1983 Fundamental and harmonic operation of millimetre-wave Gunn diodes *IEEE Trans. Microwave Theory Tech.* **MTT-31** 879–89

Johnson H R 1955 Backward wave oscillators *Proc. IRE* **43** 684–97

Kesan V P, Neikirk D P, Blakey P A, Streetman B G and Linbon T D Jr 1988 The influence of transit-time effects on the optimum design and maximum oscillation frequency of quantum well oscillators *IEEE Trans. Electron. Devices* **ED-35** 405–13

Kompfner R and Williams N T 1953 Backward wave tubes *Proc. IRE* **41** 1602–11

Kroemer H 1978 Hot-electron relaxation effects in devices *Solid-State Electron.* **21** 61–7

Mazzone A M and Rees H D 1980 Transferred-electron harmonic generators for millimetre band sources *IEE Proc.* I **127** 149–60

Palluel P and Goldberger A K 1956 The O-type carcinotron *Proc. IRE* **44** 333–45

Read W T 1958 A proposed high-frequency negative resistance diode *Bell Syst. Tech. J.* **37** 401–46

Sollner T C L G, Brown E R, Goodhue W D and Le H Q 1987 Observations of millimetre-wave oscillations from resonant tunneling diodes and some theoretical considerations of ultimate frequency limits *Appl. Phys. Lett.* **50** 332–4

Index

1/f noise, 85

Absorbers, 157
Absorptivity, 184
Anomalous phase term, 9, 16
Antenna
  area, 49
  gain, 46
  pattern, 46, 62
  solid angle, 48, 58
Antennas, 43, 56
Anti-reflection coating, 21
Aperture
  coupling efficiency, 38
  efficiency, 60
  transmission efficiency, 36
Apertures, 28, 36, 51
Attenuator, 159
Available noise power, 84
Avalanche, 198

Back-short, 69, 89
Background noise, 91
Backward-wave oscillator, 181, 186, 189
Bandwidth, 83
Barkhausen criterion, 187
Beam
  chopper, 92
  coupling, 11
  divergence angle, 58
  dump, 157

  mismatch, 177
  nodding, 95
  offset, 39
  pattern measurement, 39
  size, 9
  -splitter, 137
  throw, 23
  truncation, 34, 177
  waist plane, 10
  waist radius, 10
Bessel function, 54
Bias
  choke or tee, 100
  voltage, 98
Bismuth
  bolometer, 86
  film, 87
Black body, 181
Blazing, 20
Blooming, 20
Bolometric detector, 77, 86
Bridge arrangement, 168
Bulk mixers, 106

Cap circuit, 193
Capacitive mesh, 120
Carcinotron, 181, 186
Cassegrain
  antenna, 58
  aperture efficiency, 60
  beam pattern, 60
  far-field pattern, 62

Centre stop, 59
Characteristic impedance, 66
Chopping, 92
Chopping frequency, 93
Circuit diagrams, 156
Circular apertures, 36
Cold load, 185
Comparisons, 95
Complete record, 154
Confocal system, 180
Conversion gain, 109
Corner cube, 142
Corrugated feed, 52
Coupling efficiency, 13, 125, 228
   offset beams, 41
Cryostat, 81
Cut-off wavelength, 5, 175

Delay line, 220
Depletion region, 103
Diagonal scalar matrix, 165
Dicke switching, 95
Dielectric
   materials, 18
   sheet reflectivity, 118
   slab, 19, 72
   surface reflectivity, 19
Difference frequency, 99
Diffraction, 28
Diodes, 97, 192
Diplexer, 100, 146, 209, 222, 224
Dipole, 44
Dipole array, 122
Directionality, 46
Domains, 204
Double-balanced mixer, 211
Double-barrier diodes, 205
Double drift, 201
Double sideband, 101, 112
Down-conversion, 100
Drift region, 199

E-plane
   antennas, 56
   guide, 56
Effective area, 50
Electron coherence, 206
Emissivity, 184

Excess noise, 85

F number, 59
Fabry–Perot
   diplexer, 222
   'figure of merit', 126
   resonator, 123
Far-field patterns, 33
Fast Fourier transform, 139, 150
Feed horn, 50
Feed systems, 43
Feedback, 187
Filters, 72, 115, 123, 133
Flurogold, 133
Flurosint, 133
Focused beam waist, 24, 176
Four port system, 146
Fourier transform, 139, 184
Free-space impedance, 74
Free-standing mesh, 119
Free-standing wire grids, 140
Frequency
   conversion, 99
   discriminator, 213, 218
   error, 149
   lock loop, 210, 215
   resolution, 152
Fundamental mode, 9

Gain, 46, 49
Gallium arsenide, 104, 202
Gauss–Hermite mode patterns, 33
Gauss–Laguerre modes, 37, 54
Gaussian
   beam, 6, 124, 129, 147, 172, 221
   beam modes, 7
   modes, 16, 29
Golay detector, 86
Grids, 119
Grooves, 21
Gunn
   diodes, 201
   oscillator, 174

Half-cubes, 172
Harmonic mixer, 210
Harmonic oscillator, 203
Heat capacity, 80
Heat shielding, 81

Index

Hermite polynomials, 8
Hertzian dipole, 44
Heterodyne receivers, 97, 209
High-density polyethylene, 18
Hot electron bolometer, 86
Hot load, 185

Image frequency, 102
Impact ionization, 198
Impatt, 197
Impedance measurement, 166
Indium antimonide bolometer, 86
Indium phosphide, 202
Inductive mesh, 120
Inertial relaxation, 203
Interferometer, 136
Intermediate frequency, 99
Isolator, 170
Isotropic radiator, 46

Jerusalem cross, 120
Johnson noise, 84
Jones matrices, 161
Josephson junctions, 102
Junction devices, 102

Klystron, 181, 186, 188

Laguerre polynomials, 36
Lens
  coupling, 22
  curvature, 17
  losses and distortion, 17
  reflections, 20
  throw, 23
  waveguide, 179
Lenses, 15
Linear taper slot antenna, 57
Load, 185
  pulling, 195
  reflectivity, 69
Local oscillator, 100
Lock-in range, 215
Logic, 158
Loop
  bandwidth, 215
  gain, 214
Loss tangent, 18

Low-density polyethylene, 19
Low-dimensional structures, 205
Lower sideband, 101

Main lobe, 47
Mandril, 55
Martin–Puplett
  diplexer, 224, 231
  interferometer, 136, 140, 143
Matched load, 67
Matching diode impedance, 105
Matrices, 161
Maximum available noise power, 84
Maximum beam throw, 23, 175
Maximum signal/noise ratio, 113
Mesh, 119
Metallic mesh, 119
Michelson interferometer, 136
Mirror
  coupling, 24
  distortions, 25
  pairs, 26
  waveguide, 179
Mismatch, 70, 177
Mixer noise temperature, 112
Mixers, 97
Mode numbers, 4
Modes, 1, 65
Modulation depth, 139, 148
Multi-mode beams, 28
Multiplex advantage, 155

Negative resistance, 190
  region, 194
Noise, 82, 91
  equivalent power, 85
  measurement, 174
  rejection, 94
  suppression, 94
  temperature, 110
Non-linear devices, 97, 192
Normalized power pattern, 48
Null measurement, 95
Null reflectometer, 169
Nylon, 133
Nyquist noise, 84

Off-axis mirrors, 24

Omnidirectional antenna, 46
On-axis mirror pair, 27
One port device, 167
Open circuit, 69
Optical
  circuit diagrams, 156
  filters, 115, 123, 133
Orthogonal field, 50

Parallel beam waist, 24, 176
Paraxial beam, 7
Path difference, 139
Peak-to-valley ratio, 204
Perspex, 133
Phase, 93
  front curvature radius, 9
  lock loop, 216
  sensitive detector, 91, 212
  shifter, 159
Phonons, 85
Photoconductors, 107
Photon noise, 83, 85
Planck curve, 111, 182
P–N junctions, 102
Polarizer grids, 140, 157, 230
Polarization matrices, 161
Polarizing interferometer, 77, 136
Power gain in negative resistances, 196
Power pattern, 46
Propagation constant, 4, 68
PTFE, 18
Pyramidal horn, 51

Quality factor, Q, 193, 218
Quantum efficiency, 107

Random noise, 82
Rayleigh–Jeans region, 111
Read diode, 197
Reciprocity of antennas, 44
Rectangular
  aperture, 51
  waveguide, 2, 5, 51
Redundancy, 158
Reference signal, 93
Reflection coefficient, 68

Reflectivity, 68
Reflex klystron, 188
Refractive index, 18
Rejection ratio, 229
Resistive film, 87
Resonant
  circuit, 190
  frequency, 193
  mesh, 121
  transformer, 192
  tunnelling, 207
Resonator, 123, 174, 219
Response time, 79, 87
Responsivity, 79, 87
Ring
  diplexer, 221
  resonators, 129, 218, 230
Roof
  line, 142
  mirrors, 140

Sampling theorem, 153
Saturation, 109
Scalar feed, 52
Schottky diodes, 102, 113
Secondary
  mirror, 59
  reflectivity, 62
Semi-reflector, 72, 115
Short-circuit reflection, 67
Shot noise, 83
Shunt load, 70
Side lobes, 47
Sideband noise, 229
Sidebands, 101
Signal modulation, 93
Signal/noise ratio, 108
Single
  drift, 201
  mode, 5
  sideband, 101, 112
SIS devices, 102, 113
Sky chopping, 95
Spectrometer, 136
Spreading resistance, 104
Square-law device, 98
Synonyms, 158
System noise minimum, 113

Index

TEM modes, 56, 67
Terminations, 67, 72
Thermal
  conductivity, 80
  noise, 83
  source, 111, 181
Thin lenses, 15
Three-mirror resonator, 174
Time constant, 79
Transferred electron device, 201
Transformer, 71
Transit time, 198, 204
Transmission
  coefficient, 75
  line, 65
Transverse
  electric modes, 3
  magnetic modes, 3

Travelling-wave tube, 189
Truncation loss, 39, 177
Twin feeder, 56, 65
Two-beam interferometer, 136, 160, 183

Upper sideband, 101

Variable load, 159
Vivaldi antenna, 56

Walk-off, 128
Waveguide, 1, 43, 179
  impedance, 66
Wein displacement law, 111, 182
Wheatstone bridge, 170
Wire grids, 140, 230